T0259719

SpringerBriefs in the Mathematics of Materials

Volume 4

More information about this series at http://www.springer.com/series/13533

Koya Shimokawa • Kai Ishihara • Yasuyuki Tezuka

Topology of Polymers

 Springer

Koya Shimokawa
Department of Mathematics
Saitama University
Saitama, Japan

Kai Ishihara
Faculty of Education
Yamaguchi University
Yamaguchi, Japan

Yasuyuki Tezuka
Department of Organic
and Polymeric Materials
Tokyo Institute of Technology
Tokyo, Japan

ISSN 2365-6336 ISSN 2365-6344 (electronic)
SpringerBriefs in the Mathematics of Materials
ISBN 978-4-431-56886-5 ISBN 978-4-431-56888-9 (eBook)
https://doi.org/10.1007/978-4-431-56888-9

This Springer imprint is published by the registered company Springer Japan KK part of Springer Nature.
The registered company address is: Shiroyama Trust Tower, 4-3-1 Toranomon, Minato-ku, Tokyo 105-6005, Japan

Preface

There is a growing awareness of the importance of topology in various fields, including *polymer chemistry*. We discuss the *topology of polymers* from both mathematical and chemical viewpoints via a close collaboration of topology (Shimokawa and Ishihara) and polymer chemistry (Tezuka). We cover fundamental and selected topology topics as applied to polymers with the goal to provide novel insights revealed through the unique interaction between mathematics and polymer materials science.

We apply graph theory to analyze structures of multicyclic polymers and use terminologies to define their nomenclature. We discuss the types of multicyclic polymers, such as *spiro*, *bridged*, *fused* and *hybrid* forms, and constitutional isomers; the enumeration of multicyclic polymers is also provided. Using knot theory, we also discuss *topological isomers* and *chirality* of multicyclic polymers. In addition, we discuss the graph-theoretical and knot-theoretical properties of multicyclic polymers in practice.

With regard to polymer chemistry, we discuss the nomenclature based on the classification of mono- and poly-cycloalkanes. We then demonstrate the chemical synthesis of topologically unique multicyclic polymers via *electrostatic self-assembly and covalent fixation* protocols.

We express our sincere gratitude to Prof. Tetsuo Deguchi and Dr. Erica Uehara (Ochanomizu University) for countless invaluable discussions on this important subject. We also thank Mr. Masayuki Nakamura of Springer Japan for his continuous encouragement and support during this book project and Prof. Motoko Kotani (Tohoku University) for inviting us to contribute this monograph in SpringerBriefs in the Mathematics of Materials.

We acknowledge financial support, starting in 2014, by JSPS KAKENHI Grant-in-Aid for Scientific Research (B) (Generative Research Fields) JP26310206. This was succeeded in 2017 by the MEXT KAKENHI Grant in Aid for Scientific Research on Innovative Areas (Research in a Proposed Research Area) (Planned Research) JP17H06463.

Tokyo,   *Koya Shimokawa*
August 2019   *Kai Ishihara*
                                          Yasuyuki Tezuka

Contents

Chapter 1
Topology meets polymers: Introduction

In this chapter, we introduce polymers, long-chain molecules with diverse chemical compositions and structures. Topology can provide fundamental insights into the principle properties of polymers via their segment structures. We also present a brief description of the following chapters with respect to topological geometry and polymer chemistry.

1.1 Polymer constructions by topology insights

Polymers are ubiquitous long-chain molecules that are fundamental in biological as well as in industrial synthesis. Examples include DNA, proteins, and cellulosic compounds, as well as plastics, films, fibers, rubbers, and other common materials. There has been continuous progress in both the fundamental scientific understanding and the applications of polymers. In particular, a number of formidable breakthroughs in synthetic polymers have extended the range of structures from linear or randomly branched forms toward a variety of precisely controlled topologies, including an essential class having cyclic and multicyclic forms[1, 2]. These developments now offer opportunities to design unprecedented properties and functions via computational modeling/simulation of their forms, i.e., topologies, followed by experimental verification[3].

Because the long-chain form of polymers can be represented as a geometrical line construction, the topological (soft) geometry provides a fundamental basis to elucidate basic structural properties of flexible and randomly coiled polymers having diverse structures, in contrast to small molecules modeled on Euclidian (hard) geometry principles[4, 5]. So far, topological geometry and graph theory have successfully elucidated unique properties of DNA with cyclic, knot, and link topologies, relevant to their evolution for diverse biofunctions[6, 7]. Notably, any branching of single-stranded DNA is inherently circumvented, presumably because its principal biological function is to store and read out genetic information. In contrast, frequently encountered branching/folding structures in proteins and polypeptides are

© The Author(s), under exclusive license to Springer Japan KK 2019
K. Shimokawa et al., *Topology of Polymers*, SpringerBriefs in the Mathematics
of Materials 4, https://doi.org/10.1007/978-4-431-56888-9_1

based on controlled S-S bridging of pairs of cysteine residues found along specific locations in backbone segments. Moreover, the unique folding of linear protein or polypeptide backbone chains results in precise spatial structures crucial to their biofunctions[8]. Because folding of a linear chain into branched/cyclized forms is a geometrical transformation, topological approaches will offer valuable insights on principle factors in biological evolution. Moreover, *topological polymer chemistry* could provide conceptually new approaches to synthetic polymer chemistry and design.

Topologies of Synthetic Polymers

Linear and branched polymers

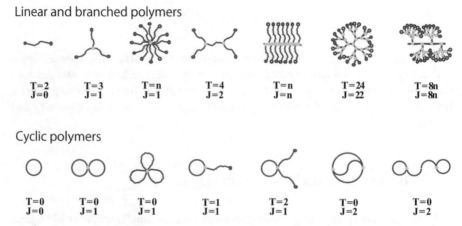

| T=2 | T=3 | T=n | T=4 | T=n | T=24 | T=8n |
| J=0 | J=1 | J=1 | J=2 | J=n | J=22 | J=8n |

Cyclic polymers

| T=0 | T=0 | T=0 | T=1 | T=2 | T=0 | T=0 |
| J=0 | J=1 | J=1 | J=1 | J=1 | J=2 | J=2 |

Fig. 1.1 Topologies of synthetic polymers with corresponding terminus (T) and junction (J) numbers as basic geometrical parameters. One cannot fully determine a structure by only using T and J. For example, two cyclic polymers in the bottom row have the same T and J. Other concepts for structure characterization are introduced in Chapter 2 and Chapter 3.

A graph presentation of prototypical polymer branched or cyclic constructions are depicted at the top and bottom, respectively, of Fig. 1.1. From left to right at the top of Fig. 1.1 are shown simple linear, three-armed, and n-armed "star" polymers, "comb-shaped" polymers (polymacromonomers)[9], "dendrimers,"[10], and "dendron-jacketed" polymers[11]. At the bottom of Fig. 1.1 are shown, from left to right, a simple ring, and "8-shaped," "trefoil-shaped," "tadpole-shaped," "twin-tailed tadpole-shaped," "θ-shaped," and "manacle (handcuff)-shaped" polymers[12, 13]. The topologically intriguing polymers in Fig. 1.1 have been constructed because of the remarkable progress in synthetic polymer chemistry.

Also given in Fig. 1.1, are the terminus (T, chain end) and the junction (J, branching point) numbers, which are the two principal geometrical parameters of each construction. They are taken as invariant and crucial when distinguishing constructions from topological viewpoints. Additionally, the total number of branches at each junction and the connectivity of each junction are considered invariant geometric parameters. Whereas, geometrical parameters such as the distance between two adjacent junctions and that between the junction and the terminus, are variable geometric parameters in flexible long-chain polymers capable of assuming a random coils. Notably, flexible randomly coiled polymer segments do not conform to Euclidian geometry where the distance between two adjacent junctions and that between the junction and terminus are invariant.

"Dendrimer" and "dendron-jacketed" polymer constructions are in contrast to flexible polymers because the distance between two adjacent junctions and that between the junction and the terminus are postulated as short and constant. Thus, their geometrical nature is characterized as "Euclidian" rather than topological, and corresponds to their stiff, shape-persistent characteristics in a gradient of the segment density[11]. The "comb-shaped" construction is partially topological because the branch chains are flexible, while the distance between two junctions along the backbone is short and constant. This is in contrast to the "dendron-jacketed" polymers having shape-persistent branch units.

For the branched polymers in Fig. 1.1, increasingly complex structures have increasing terminus and junction numbers. For the cyclic polymers, however, diverse constructions are produced with small numbers. Ring construction also has the minimal parameter numbers of $T = 0$ and $J = 0$, while linear construction has $T = 2$ and $J = 0$. Hence, from the topological geometry viewpoint, the ring form is the simplest construction rather than the linear one, which has traditionally been regarded as the most primitive construction because nearly all available polymer substrates, especially synthetic ones, are basically linear when produced via "polymerization" of monomer substrates and unidirectional propagation/linking. Recently, the production of ring/cyclic polymers has become a common practice, and a renewed theory of polymers based on geometrical/topological considerations will hopefully be formulated.

1.2 From topological analyses to constructing graph-structure polymers

In this monograph, principal concepts and basic practices of *topological polymer chemistry* are discussed in terms of topological geometry and polymer chemistry.

In Chapter 2, we discuss theoretical methods of analyzing structures of multicyclic polymers. We first define a polymer graph that represents the structure. Then, we introduce a systematic notation of polymer structures using their graphs, and reveal the enumeration process of multicyclic polymers using the notation method. Finally, we introduce the folding operation and characterize which graph can be obtained from a line graph via folding.

In Chapter 3, we discuss chemistry-based hierarchical classification in which nonlinear, cyclic, and branched polymers are classified upon the molecular graph presentation of cycloalkanes. By using graph presentations of alkane isomers (C_nH_{2n+2}), and a series of mono- and poly-cycloalkanes (e.g., C_nH_{2n}, C_nH_{2n-2}), the hierarchical classification from simple to complex branched and cyclic polymer topologies is introduced. This allows grouping into different main classes and subclasses. A systematic notation protocol for nonlinear polymer topologies is also proposed based on the principal geometrical parameters T and J [12].

In Chapters 4, we consider types of graphs and define spiro-, bridged-, fused-, and hybrid graph constructions. The characterization of diverse graph constructions could suggest synthetic methods for polymers having those graph structures.

In Chapter 5, we consider the concept of topological isomerism involving multicyclic polymers. This is with respect to knots, links, and spatial graphs based on standard knot theories and low-dimensional topology conjectures.

In Chapter 6, we discuss the geometrical/topological relationship in practice, with a particular emphasis on isomerism and its geometrical graph transformation. A unique concept of *topological isomerism* is introduced that contrasts with conventional constitutional and stereoisomerism in small molecules. These conceptual/theoretical considerations could provide rational and effective synthetic pathways of complex polymers having cyclic and multicyclic topologies.

In Chapter 7, the *electrostatic self-assembly and covalent fixation* (ESA-CF) protocol is demonstrated by employing specifically designed *telechelic* precursors with selected cyclic ammonium salt end groups, to construct complex polymer topologies, including cyclic and multicyclic polymer units[13]. A wide range of complex, systematically classified multicyclic polymers having specific *spiro-*, *bridged-*, *fused-*, and *hybrid*-forms have been constructed via ESA-CF in conjunction with effective linking chemistries such as alkyne–azide addition (*click*) and an olefin metathesis (*clip*). Finally, in a remarkable showcase of *topological polymer chemistry*, the construction of a topologically significant macromolecular $K_{3,3}$ graph topology is described[14].

In Chapter 8, a conclusion is presented.

References

1. N. Hadjichristidis, A. Hirao, Y. Tezuka, and F. Du Prez, Eds, Complex Macromolecular Architectures, (Wiley, Singapore, 2011).
2. A. D. Schlüter, C. J. Hawker, and J. Sakamoto, Eds., Synthesis of Polymers, New Structures and Methods (Wiley-VCH, Weinheim, 2012).
3. Y. Tezuka, Ed., Topological Polymer Chemistry: Progress of cyclic polymers in syntheses, properties and functions (World Scientific, Singapore, 2013).
4. D. M. Walba, Tetrahedron, **41**, 3161 (1985).
5. E. Flapan, When Topology Meets Chemistry: A Topological Look at Molecular Chirality (Cambridge University Press, Cambridge, 2000).
6. N. C. Seeman, Annu. Rev. Biochem., **79**, 65 (2010).
7. K. Shimokawa, K. Ishihara, I. Grainge, D. J. Sherratt, and M. Vazquez, Proc. Nat. Acad. Sci., USA, **110**, 20906 (2013).
8. D. J. Craik, Nature Chem., **4**, 600 (2012).
9. K. Ito and S. Kawauchi, Adv. Polym. Sci., **142**, 129 (1999).
10. A. D. Schlüter and J. P. Rabe, Angew. Chem. Int. Ed., **39**, 864 (2000).
11. M. Fischer and F. Vögtle, Angew. Chem. Int. Ed., **38**, 884 (1999).

12. Y. Tezuka and H. Oike, J. Am. Chem. Soc., **123**, 11570 (2001).
13. Y. Tezuka, Acc. Chem. Res., **50**, 2661 (2017).
14. T. Suzuki, T. Yamamoto, and Y. Tezuka, J. Am. Chem. Soc., **136**, 10148 (2014).

Chapter 2
Graph theory analyses of polymers

In this chapter, we introduce graph theory for analyzing structures of multicyclic polymers. An essential graph theory reference is [1].

In Section 2.1, we introduce basic concepts of graphs in which we define a polymer graph that represents a polymer structure. In Section 2.2, we introduce a systematic notation of polymer structure by using their graphs. This nomenclature and its application are the main themes of this chapter. In Section 2.3, we enumerate multicyclic polymers using the notation. Figures of graphs with rank 4 at most are discussed here. In Section 2.4, we discuss an application of L. Euler's solution of the Königsberg bridge problem. We introduce the folding operation and characterize what graph can be obtained from a simple linear graph by folding. As an application, we discuss the characterization of multicyclic polymers obtained from a simple linear graph by folding.

There are many references on applications of graph theory to the enumeration of isomeric structures. See, for example, [2, 3, 4, 5, 6, 7, 8, 9, 10, 11, 12]. See also [13, 14].

2.1 Graphs

The structures of polymers can expressed and studied as "graphs". In this section, we introduce mathematical concepts of graphs.

2.1.1 Definition of graphs

Definition 2.1. A *graph* is an object that consists of vertices and edges, where each vertex is a point, and each edge is a segment with endpoints that are vertices. See Fig. 2.1. A graph G is denoted by the pair (V, E), where V and E are sets of vertices and edges, respectively.

© The Author(s), under exclusive license to Springer Japan KK 2019
K. Shimokawa et al., *Topology of Polymers*, SpringerBriefs in the Mathematics of Materials 4, https://doi.org/10.1007/978-4-431-56888-9_2

Let $G = (V,E)$ and $G' = (V',E')$ be graphs. Graph G' is a *subgraph* of G if $V' \subset V$ and $E' \subset E$ hold. A graph G is *finite* if V and E are both finite sets. In this monograph, we consider only finite graphs. An edge is a *loop* if two endpoints coincide. A *degree* (or *valency*) of a vertex v, denoted by $d(v)$, is the sum of non-loop edges incident to v and twice the number of loops incident to v. For example, the degree of the vertex v_1 in Fig. 2.1 is 5, *i.e.* $d(v_1) = 5$. The terminus and junction are vertices with degree 1 and degree at least 3, respectively. A terminus is sometimes a *pendant vertex* or a *leaf*, and the unique edge incident to the terminus is a pendant edge.

For an edge e whose endpoints are vertices v_1 and v_2, the boundary of e is the set $\{v_1, v_2\}$ and denotes it by $\partial e = \{v_1, v_2\}$. In this case, v_1 and v_2 are *adjacent*. A graph $G = (V,E)$ is a *path* of length k if $V = \{v_0, \cdots, v_k\}$ and $E = \{e_1, \cdots, e_k\}$, such that $\partial e_i = \{v_{i-1}, v_i\}$ and all the vertices v_i's are distinct.

A graph G is *connected* if there is a path between any pair of vertices. Otherwise, the graph is *disconnected*. A closed path, *i.e.*, $v_0 = v_k$, is a *cycle*. A graph without cycles is a *forest*. In particular, a connected forest is a *tree*. A *maximal tree* of a graph G is a tree subgraph of G containing the maximal number of edges among all tree subgraphs.

The *rank* of a graph is the minimal number of edges needed to be removed from G to make it into a forest; it is denoted by $r(G)$. Throughout this chapter, n, m, ℓ, r are defined as the following.

1. $n = |V|$ (the number of elements of V).
2. $m = |E|$ (the number of elements of E).
3. $\ell = |L|$ (the number of elements of L), where L is the set of loops of G.
4. $r = r(G)$ (the rank of G).

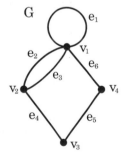

Fig. 2.1 Example of a graph $G = (V,E)$, where $V = \{v_1, v_2, v_3, v_4\}$ is the set of vertices and $E = \{e_1, e_2, e_3, e_4, e_5, e_6\}$ is the set of edges. The *valency* of v_1 is 5 and this graph has *rank* 3 and $T = 0$.

Proposition 2.1. *For any connected finite graph G, the following holds:*

$$r = m - n + 1$$

Proof. Let T_G be a maximal tree of G. Note where T_G contains all the vertices of G.

By applying mathematical induction, we can show that T_G has $n-1$ edges. Each edge of G that is not contained in T corresponds to a cycle of G. Hence, the rank of G is $m-(n-1)=m-n+1$.

If two graphs have the same number of vertices connected in the same way, they have the same topological construction.

Definition 2.2. Two graphs $G_1 = (V_1, E_1)$ and $G_2 = (V_2, E_2)$ are *isomorphic* if there are two bijections $\varphi : V_1 \to V_2$ and $\psi : E_1 \to E_2$, such that $\partial \psi(e) = \varphi(\partial e)$ for any $e \in E_1$.

Example 2.1. The graph on the left in Fig. 2.2 is a θ *(theta) graph* and the graph on the right is a *manacle graph* or a *handcuff graph*. Although each graph contains two vertices and three edges, they are not isomorphic.

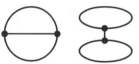

Fig. 2.2 Example of graphs. On the left is a θ-graph, and on the right is a manacle (handcuff) graph.

2.1.2 Graphs associated with molecules and polymers

A *molecular graph* [5, 6] has atom vertices and edges that are chemical bonds. A *hydrogen-depleted molecular graph, i.e.*, is a molecular graph without hydrogen vertices. The structure of alkanes is discussed in Chapter 3 by using this molecular graph. For example, constitutional isomers can be obtained with a pair of non-isomorphic graphs having the same number of vertices.

To study polymer structures, we use the following coarse-grained representation. Here, a *vertex* is a branch point of a polymer that is a collection of atoms, and an *edge* is a linear structure that is a chain of bonds between two branched points. For multicyclic polymers, termini vertices and vertices with degree two are often deleted. In this setting, the degrees of vertices are at least 3. See Fig. 2.4, where a vertex corresponds to a benzene ring. A graph so constructed is *graph (structure) associated with a molecule or a polymer*, or a *polymer graph*.

Let G be the graph associated with a polymer. A graph or polymer is *monocyclic* (or *dicyclic, tricyclic, tetracyclic* or *pentacyclic*) if $r(G) = 1$ (or $2, 3, 4$ or 5, respectively).

Fig. 2.3 An example of a molecular graph. Vertices are atoms and edges are chemical bonds. A simplified version drops vertices with degree 2.

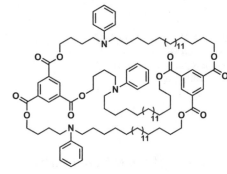

Fig. 2.4 Example of a molecular graph associated with a polymer. It is a θ-graph with rank 2. Hence, the polymer is dicyclic.

2.1.3 Adjacency matrix

In this subsection, we introduce the adjacency matrix of a graph. Let $A = (a_{ij})$ denote a matrix with a_{ij} as the (i,j) entry.

Definition 2.3 (Adjacency matrix). For a graph $G = (V,E)$ with $V = \{v_1,\ldots,v_n\}$, let a_{ij} be the number of edges of G connecting v_i and v_j for $i,j \in \{1,\ldots,n\}$. The *adjacency matrix* A_G of G is a $n \times n$ matrix with (i,j) entry a_{ij}. Note that the diagonal entry $a_{i,i}$ corresponds to the number of loops incident to v_i.

Example 2.2. The adjacency matrix of the graph in Fig. 2.1 is:

$$\begin{pmatrix} 1 & 2 & 0 & 1 \\ 2 & 0 & 1 & 0 \\ 0 & 1 & 0 & 1 \\ 1 & 0 & 1 & 0 \end{pmatrix}$$

Remark 2.1. By definition, an adjacency matrix is *symmetric*, i.e., $a_{ji} = a_{ij}$ for each $i, j \in \{1, \ldots, n\}$. Conversely, for any symmetric matrix A without negative entries, a graph having the adjacency matrix A is uniquely determined.

We can easily check that the following equations hold.

Proposition 2.2. *Let $A_G = (a_{ij})$ be the adjacency matrix of graph G. Then the following hold:*

1. $d(v_i) = a_{ii} + \sum_{j=1}^{n} a_{ij}$, i.e., the sum of the entries of the i-th row and a_{ii} is the degree of v_i.

2. $m = \dfrac{1}{2} \sum_{i=1}^{n} d(v_i) = \sum_{1 \le i \le j \le n} a_{ij}$, i.e., the sum of all upper right entries including diagonal entries is the number of edges.

2.1.4 Complete graphs K_n and complete bipartite graphs K_{n_1, n_2}

In this subsection, we introduce the complete graph K_n and the complete bipartite graph K_{n_1, n_2}.

Definition 2.4. A *complete graph K_n* consists of n vertices, where each pair of vertices is connected by a unique edge.

The number of edges in K_n is $\frac{n(n-1)}{2}$.

Definition 2.5. A graph $G = (V, E)$ is a *bipartite graph* if V admits a partition into two subsets V_1 and V_2, such that each edge in E connects a vertex in V_1 with a vertex in V_2. If each pair of vertices in V_1 and V_2 are adjacent, then it is a *complete bipartite graph*. If V_1 contains n_1 vertices and V_2 contains n_2 vertices, it is K_{n_1, n_2}. The number of edges in K_{n_1, n_2} is $n_1 n_2$.

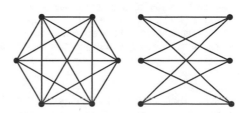

Fig. 2.5 A complete graph K_6 (left) and a complete bipartite graph $K_{3,3}$ (right).

Proposition 2.3. *1. $K_{3,3}$ does not contain a pair of disjoint cycles.*
2. $K_{3,3}$ contains 6 cycles of length 6 and 9 cycles of length 4.

2.2 Systematic notation of polymers

In this section, we present a nomenclature of polymers using associated graphs. It is basically the same as that reported previously [17]. We also show that the rank can be easily calculated by its notation.

First, we define the degree sequence that is used in the nomenclature.

Definition 2.6. Let $V = \{v_1, \cdots, v_n\}$. Consider the degree of all vertices and arrange them in non-increasing order. The sequence $(d_1, \cdots, d_n)$ is *the degree sequence* of G. To simplify the sequence, sometimes we indicate a (sub)sequence $d_i, \cdots, d_i$ of length n_i as $d_i^{n_i}$.

The graph is sometimes uniquely determined from the degree sequence. For example, an m-armed star graph is a unique graph with degree sequence $(m, 1, \ldots, 1) = (m, 1^m)$. Two graphs can have the same degree sequence even if they are not isomorphic. For example, the θ-graph and the manacle graph both have $(3,3) = (3^2)$, even though the θ-graph has no loop and the manacle graph has two loops. They can be distinguished by the number of loops. There are graphs with the same degree sequence and the same number of loops. To distinguish these graphs, we number them in a certain way. See Fig. 2.6.

Now we define a nomenclature of graphs.

Definition 2.7 (A nomenclature of graphs). A graph with the degree sequence $(d_1, \cdots, d_n)$ with ℓ loops is $(d_1, \cdots, d_n)_k^\ell$, where k is the numbering. We omit ℓ when $\ell = 0$.

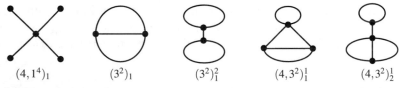

$(4, 1^4)_1$ $\qquad$ $(3^2)_1$ $\qquad$ $(3^2)_1^2$ $\qquad$ $(4, 3^2)_1^1$ $\qquad$ $(4, 3^2)_2^1$

Fig. 2.6 Examples of graph nomenclature. A 4-armed star graph is a unique graph with degree sequence $(4, 1^4)$. There are exactly two graphs (a θ graph and a manacle graph) with the same degree sequence (3^2), The θ graph does not contain a loop and the manacle graph contains two loops. There are exactly two graphs with the same degree sequence $(4, 3^2)$ with one loop.

The adjacency matrix detects most of the nomenclature except the numbering k. Clearly, a graph with the nomenclature $(d_1, \cdots, d_n)_k^\ell$ has n vertices. In fact, the number of edges and the rank are also determined directly from the nomenclature.

Proposition 2.4 (The adjacency matrix and the nomenclature). *Let $A = (a_{ij})$ be the adjacency matrix of G with the nomenclature $(d_1, \cdots, d_n)_k^\ell$. Then the following hold:*

1. $\ell = \sum_{i=1}^{n} a_{ii}$.

2. $d_{\sigma(i)} = a_{ii} + \sum_{i=1}^{n} a_{ij}$ *for each* $i \in \{1, \cdots, n\}$, *where* $\sigma \in S_n$ *is some permutation.*

3. $m = \dfrac{1}{2} \sum_{i=1}^{n} d_i$, *where m is the number of edges.*

4. $r = \dfrac{1}{2} \sum_{i=1}^{n} (d_i - 2) + 1$, *where r is the rank.*

Proof. A diagonal entry a_{ii} of A is the number of loops incident to the i-th vertex. Thus, the sum of all entries is equal to ℓ, which is the first equation. The second and third equations follow from Proposition 2.2. Substituting the third equation into the equation in Proposition 2.1, yields the fourth equation:

$$r = m - n + 1 = \frac{1}{2} \sum_{i=1}^{n} d_i - n + 1 = \frac{1}{2} \sum_{i=1}^{n} (d_i - 2) + 1$$

Example 2.3. The ranks of graphs $(4, 1^4)_1$, $(3^2)_1$, $(3^2)_1^2$, $(4, 3^2)_1^1$ and $(4, 3^2)_2^1$ in Fig. 2.6 are 0, 2, 2, 3 and 3, respectively.

2.3 Enumeration of graphs associated with multicyclic polymers

In this section, we demonstrate a method for enumerating ring polymers. We include the lists of dicyclic, tricyclic and tetracyclic graphs in Theorems 2.1, 2.2 and 2.3, respectively. The same argument appears in [18].

In this enumeration, we focus our attention on graphs with degree at least 3. For multicyclic polymers, we often delete vertices with degree 1 and the edge incident to it, i.e., we assume T= 0. We also delete vertices with degree 2 because it does not change the polymer structure. Thus, we assume that every vertex is a junction.

Let us consider graphs with degree sequence $(d_1, \cdots, d_n)$ and rank $r \geq 2$, where $d_i \geq 3$ for each i. From Proposition 2.4, we have:

$$\sum_{i=1}^{n} (d_i - 2) = 2r - 2.$$

Thus, the condition $d_i \geq 3$ implies $n \leq 2r - 2$. From these inequalities, there are only a limited number of possible degree sequences for each r. Because any symmetric matrix uniquely determines a graph having it as the adjacency matrix, we enumerate symmetric matrices (a_{ij}) so that $a_{ij} \geq 0$ and $a_{ii} + \sum a_{ij} = d_i$. The possibility of such symmetric matrices are also limited for each degree sequence. Thus, there are a finite number graphs for each r.

First, we enumerate dicyclic graphs.

Theorem 2.1. *There are exactly* 3 *dicyclic graphs with degree at least* 3 *at each vertex, as in Fig. 2.7. Namely* $(3^2)_1$, $(4)^2_1$ *and* $(3^2)^2_1$.

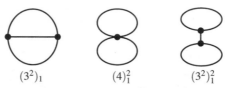

$$(3^2)_1 \qquad\qquad (4)^2_1 \qquad\qquad (3^2)^2_1$$

Fig. 2.7 There are exactly 3 dicyclic graphs (i.e., graphs with rank 2) with degree at least 3 at each vertex.

Here, we give a proof of Theorem 2.1.

Proof. From the condition $\sum_{i=1}^{n}(d_i - 2) = 2$ and $d_i \geq 3$, either $n = 1$ and $d_1 = 4$ or $n = 2$ and $(d_1, d_2) = (3, 3)$. In the former case, every edge is a loop because the graph has only one vertex; thus the graph has to be $(4)^2_1$ in Fig. 2.7. In the latter case, we arrange the numbers a_{11}, $a_{12} = a_{21}$, and a_{22} so that $2a_{11} + a_{12} = d_1 = 3$ and $a_{21} + 2a_{22} = d_2 = 3$ to obtain the adjacency matrix. However, $a_{12} = 0$ implies that there is no edge between the two vertices; so the graph is disconnected. Thus, the possible adjacency matrices are the following two.

$$\begin{pmatrix} 0 & 3 \\ 3 & 0 \end{pmatrix} \qquad \begin{pmatrix} 1 & 1 \\ 1 & 1 \end{pmatrix}$$

They represent $(3^2)_1$ and $(3^2)^2_1$ in Fig 2.7, respectively. This completes the proof of Theorem 2.1.

Next, we enumerate tricyclic graphs.

Theorem 2.2. *There are exactly* 15 *tricyclic graphs with degree at least* 3 *at each vertex as in Fig. 2.8.*

This characterization is consistent with [15, 16]. In particular, the four graphs $(3^4)_1$, $(3^4)_2$, $(4, 3^2)_1$ and $(4^2)_1$ that have no loops, are α, β, γ and δ graphs, respectively [19].

Here we give an outline of the proof for Theorem 2.2.

Proof. Basically, this is the same argument as that for the proof of Theorem 2.1. From the condition $\sum_{i=1}^{n}(d_i - 2) = 4$ and $d_i \geq 3$, the possible degree sequences are the following five:

$$(6), (5, 3), (4, 4), (4, 3, 3), (3, 3, 3, 3)$$

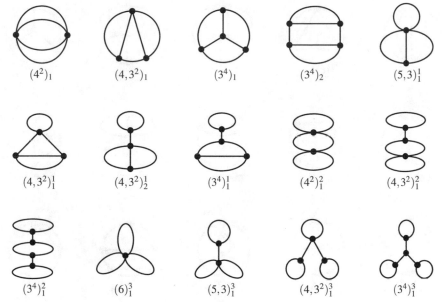

$(4^2)_1$ $(4,3^2)_1$ $(3^4)_1$ $(3^4)_2$ $(5,3)_1^1$

$(4,3^2)_1^1$ $(4,3^2)_2^1$ $(3^4)_1^1$ $(4^2)_1^2$ $(4,3^2)_1^2$

$(3^4)_1^2$ $(6)_1^3$ $(5,3)_1^3$ $(4,3^2)_1^3$ $(3^4)_1^3$

Fig. 2.8 There are exactly 15 tricyclic graphs (i.e., graphs with rank 3) with degree at least 3 at each vertex.

For each degree sequence, we arrange the numbers a_{ij} to obtain the adjacency matrix. Consider the degree sequence $(4,3,3)$ as an example. The possible adjacency matrices are the following five up to permutations:

$$\begin{pmatrix} 0 & 2 & 2 \\ 2 & 0 & 1 \\ 2 & 1 & 0 \end{pmatrix} \quad \begin{pmatrix} 1 & 1 & 1 \\ 1 & 0 & 2 \\ 1 & 2 & 0 \end{pmatrix} \quad \begin{pmatrix} 0 & 3 & 0 \\ 3 & 0 & 1 \\ 0 & 1 & 1 \end{pmatrix} \quad \begin{pmatrix} 1 & 2 & 0 \\ 2 & 0 & 1 \\ 0 & 1 & 1 \end{pmatrix} \quad \begin{pmatrix} 1 & 1 & 1 \\ 1 & 1 & 0 \\ 1 & 0 & 1 \end{pmatrix}$$

These represent $(4,3^2)_1$, $(4,3^2)_1^1$, $(4,3^2)_2^1$, $(4,3^2)_1^2$ and $(4,3^2)_1^3$ in Fig 2.8, respectively. Continuing this argument, we complete the proof of Theorem 2.2.

Next, we enumerate tetracyclic graphs

Theorem 2.3. *[18] There are exactly* 111 *tetracyclic graphs with degree at least* 3 *at each vertex as in Fig. 2.9.*

The famous Königsberg bridge graph has rank 4 and its nomenclature is $(5,3,3,3)_1 = (5,3^3)_1$. The $K_{3,3}$ graph is $(3,3,3,3,3,3)_1 = (3^6)_1$.

Here we give an outline of the proof of Theorem 2.3 that is based on the arguments in the proofs of Theorems 2.1 and 2.2.

Proof. From the condition $\sum_{i=1}^{n}(d_i - 2) = 6$ and $d_i \geq 3$, the possible degree sequences are the following:

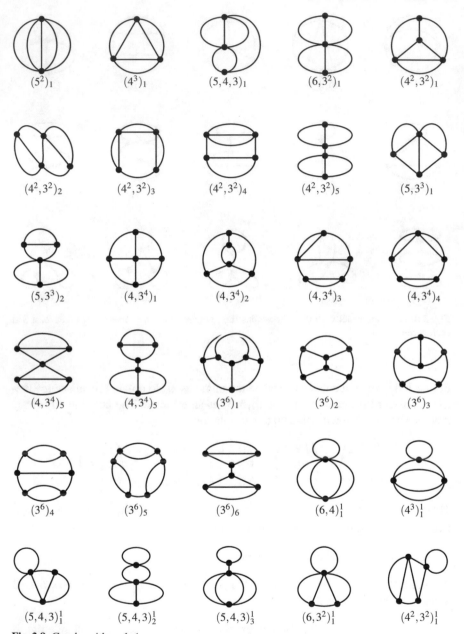

Fig. 2.9 Graphs with rank 4.

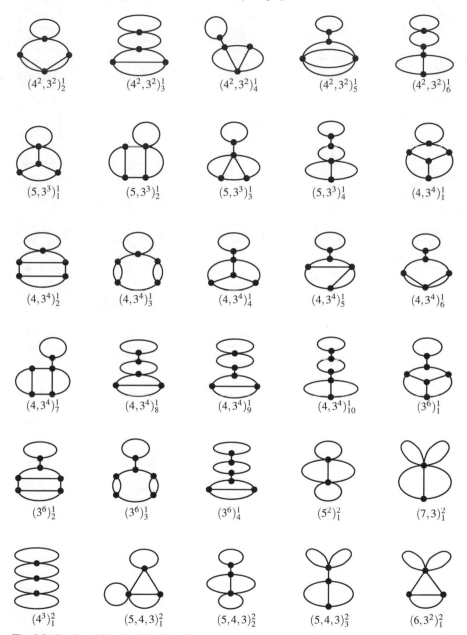

$(4^2,3^2)^1_2$ $(4^2,3^2)^1_3$ $(4^2,3^2)^1_4$ $(4^2,3^2)^1_5$ $(4^2,3^2)^1_6$

$(5,3^3)^1_1$ $(5,3^3)^1_2$ $(5,3^3)^1_3$ $(5,3^3)^1_4$ $(4,3^4)^1_1$

$(4,3^4)^1_2$ $(4,3^4)^1_3$ $(4,3^4)^1_4$ $(4,3^4)^1_5$ $(4,3^4)^1_6$

$(4,3^4)^1_7$ $(4,3^4)^1_8$ $(4,3^4)^1_9$ $(4,3^4)^1_{10}$ $(3^6)^1_1$

$(3^6)^1_2$ $(3^6)^1_3$ $(3^6)^1_4$ $(5^2)^2_1$ $(7,3)^2_1$

$(4^3)^2_1$ $(5,4,3)^2_1$ $(5,4,3)^2_2$ $(5,4,3)^2_3$ $(6,3^2)^2_1$

Fig. 2.9 Graphs with rank 4 (continued).

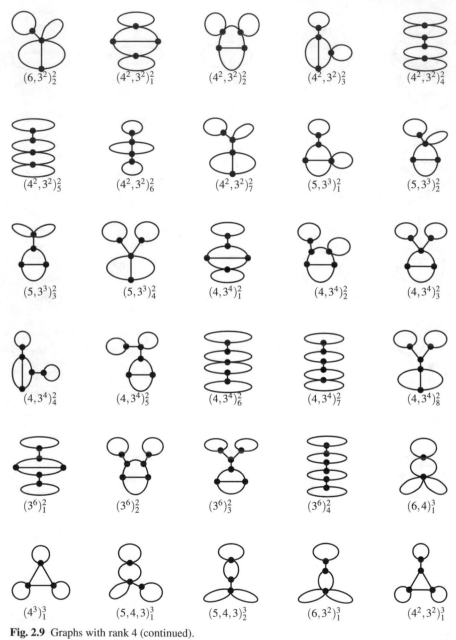

$(6,3^2)^2_2$ $(4^2,3^2)^2_1$ $(4^2,3^2)^2_2$ $(4^2,3^2)^2_3$ $(4^2,3^2)^2_4$

$(4^2,3^2)^2_5$ $(4^2,3^2)^2_6$ $(4^2,3^2)^2_7$ $(5,3^3)^2_1$ $(5,3^3)^2_2$

$(5,3^3)^2_3$ $(5,3^3)^2_4$ $(4,3^4)^2_1$ $(4,3^4)^2_2$ $(4,3^4)^2_3$

$(4,3^4)^2_4$ $(4,3^4)^2_5$ $(4,3^4)^2_6$ $(4,3^4)^2_7$ $(4,3^4)^2_8$

$(3^6)^2_1$ $(3^6)^2_2$ $(3^6)^2_3$ $(3^6)^2_4$ $(6,4)^3_1$

$(4^3)^3_1$ $(5,4,3)^3_1$ $(5,4,3)^3_2$ $(6,3^2)^3_1$ $(4^2,3^2)^3_1$

Fig. 2.9 Graphs with rank 4 (continued).

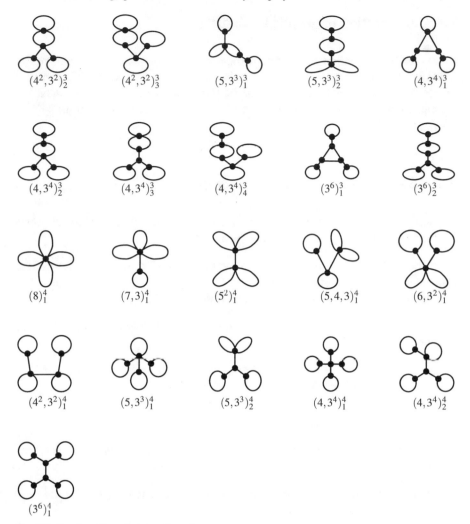

Fig. 2.9 Graphs with rank 4 (continued).

$$(8), (7,3), (6,4), (5^2), (6,3^2), (5,4,3), (5,3^3), (4^2,3^2), (4,3^4), (3^6)$$

We enumerate all possible adjacency matrices for each degree sequence. Continuing this argument, we complete the proof of Theorem 2.3.

2.4 Construction of graphs by folding

In this section, we introduce the *folding* operation of a graph. We will characterize multicyclic graphs obtained from a simple linear graph by folding. We can then discuss which multicyclic polymers can be synthesized from linear polymers. For example, a polymer with the $K_{3,3}$ graph structure cannot be obtained from a linear polymer by folding.

2.4.1 Eulerian and semi-Eulerian graphs

We start with the definitions of Eulerian and semi-Eulerian graphs. These concepts are related to the Königsberg bridge problem[20].

Definition 2.8. Let G be a graph. A path or a cycle of G is *Eulerian* if it traverses each edge of G exactly once. A graph G is *Eulerian* if G contains a Eulerian cycle. A graph G is *semi-Eulerian* if G contains a Eulerian path.

The Königsberg bridge graph is not semi-Eulerian. The next proposition is due to Euler.

Proposition 2.5. *[20, 1] A connected graph G with nomenclature $(d_1, \cdots, d_n)_k^\ell$ is semi-Eulerian if and only if one of the following holds:*

1. Each d_i is even.
2. There are exactly two odd d_is.

In case 1, G is Eulerian. In case 2, G is not Eulerian and the starting point (and end point) of the Eulerian path is a vertex with an odd degree.

Because the Königsberg bridge graph has the nomenclature $(5,3,3,3)_1$, it is neither Eulerian nor semi-Eulerian.

2.4.2 Construction of multicyclic polymers by folding linear polymers

In this subsection we define the *folding* operation.

Definition 2.9. A graph G is *obtained from G' by a folding* if G is obtained by identifying several vertices of G'. Here, three or more vertices may be identified to make a new vertex. See Fig 2.10. A graph G is *obtained from G' by a simple folding* if G is obtained from G' by identifying pairs of vertices in G'. Here, we only consider operations as in Fig. 2.11, *i.e.*, identifying a pair of vertices having degree one or two.

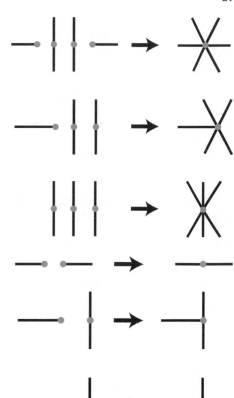

Fig. 2.10 Folding operations.

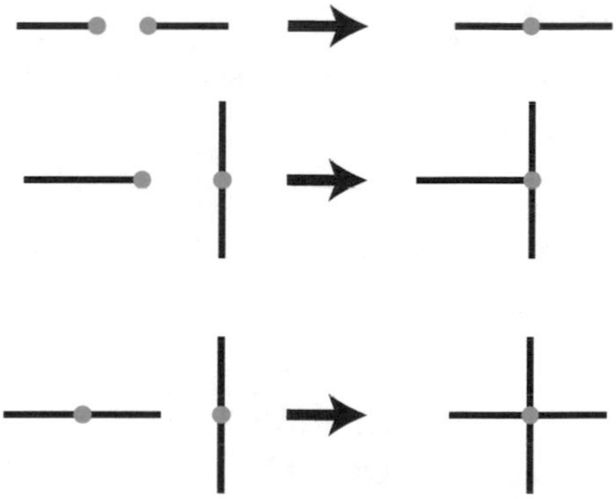

Fig. 2.11 Simple folding
operations.

Definition 2.10. If G_1 and G_2 are obtained from a graph G by folding, then G_1 and G_2 are *folding isomers*.

Note that folding isomers are examples of constitutional isomers. Next, we define a simple linear graph.

Definition 2.11. A *simple linear graph* with length m is a graph consisting of only one path with m edges, and is denoted by L_m.

The simple linear graph L_m of length m is a unique graph with the degree sequence $(2,\ldots,2,1,1) = (2^{m-1},1^2)$; hence its nomenclature is $(2^{m-1},1^2)_1$.

Note that if a graph G is obtained from L_m by folding, then the number of edges in G is m. When all vertices are identified by the folding $G = (2m)_1^m$, and the rank of G becomes m, then that is the largest of all graphs obtained from L_m by folding. Furthermore, the degree of each vertex of a graph that is obtained from L_m by simple folding is at most 4. By a simple application of Proposition 2.5, we characterize multicyclic graphs obtained from a simple linear graph by folding. By folding, the simple linear graph becomes an Eulerian path in the resulting graph.

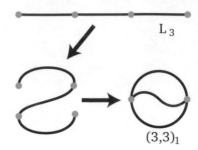

Fig. 2.12 The graph $(3,3)_1$ is obtained from L_3 by a simple folding.

$(3,3)_1$

Theorem 2.4. *A graph G is obtained from a simple linear graph L_m by a folding of some m if and only if G is semi-Eulerian.*

Theorem 2.5. *Let G be a connected circular polymer graph with the nomenclature $(d_1,\cdots,d_n)_k^\ell$.*

1. G can be obtained from L_m by folding if and only if G is semi-Eulerian and
$$\sum_{i=1}^{n} d_i = 2m.$$

2. G can be obtained from L_m by simple folding if and only if $\sum_{i=1}^{n} d_i = 2m$ and one of the following holds:

a. $(d_1,\cdots,d_n) = (2^n)$ or $(4^{n_1},2^{n_2})$.
b. $(d_1,\cdots,d_n) = (4^{n_1-1},3^2,2^{n_2-1})$.
c. $(d_1,\cdots,d_n) = (4^{n_1-1},3,2^{n_2-1},1)$.
d. $(d_1,\cdots,d_n) = (4^{n_1-1},2^{n_2-1},1^2)$.

Here, n_1 and n_2 are positive integers with $n_1+n_2=n$.

Proof. Because folding (or simple folding) does not change the number of edges, the resulting graph also has m edges. Hence, we have $\sum_{i=1}^{n} d_i = 2m$. By folding (or simple folding), L_m becomes a Eulerian graph in G. By Proposition 2.5, the theorem follows.

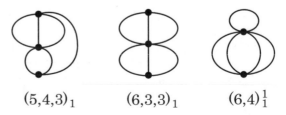

Fig. 2.13 Graphs obtained from L_m by folding.

$(5,4,3)_1$ $(6,3,3)_1$ $(6,4)_1^1$

By an application of Theorem 2.5, we can show that $K_{3,3} = (3,3,3,3,3,3)_1 = (3^6)_1$ cannot be prepared by folding a simple linear graph. Hence, we need an alternative approach for synthesizing a polymer with the $K_{3,3}$ structure. We discuss the chemical synthesis of such a polymer in Chapter 7.

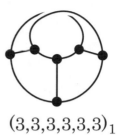

$$(3,3,3,3,3,3)_1$$

Fig. 2.14 $K_{3,3}$ cannot be obtained from L_m by folding.

Example 2.4. A graph $G \neq L_2$ is obtained from L_2 by folding if and only if $G = (2,2)_1$, $(3,1)_1^1$ or $(4)_1^2$. See Fig. 2.15. A graph $G \neq L_3$ is obtained from L_3 by folding if and only if $G = (2,2,2)_1$, $(3,2,1)_1$, $(3,2,1)_1^1$, $(4,1,1)_1^1$, $(3,3)_1$, $(3,3)_1^2$, $(4,2)_1^1$, $(5,1)_1^2$ or $(6)_1^3$. Moreover those graphs, except $(5,1)_1^2$ and $(6)_1^3$, arise from simple folding. See Fig. 2.16. A connected polymer graph $G = (4,4)_1, (5,3)_1^1$, $(4,4)_1^2, (5,3)_1^3, (8)_1^4$ can be obtained from L_4 by folding. See Fig. 2.17.

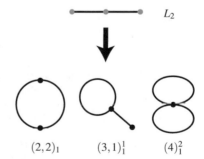

$$L_2$$

$(2,2)_1$ $(3,1)_1^1$ $(4)_1^2$

Fig. 2.15 Graphs $(2,2)_1$, $(3,1)_1^1$, and $(4)_1^2$ can be obtained from L_2 by folding.

References

1. R. Diestel, Graph Theory, 5th edition, Graduate Texts in Mathematics, Springer, 2017.
2. A.T. Balaban, ReV. Roum. Chim. (1973), **18**, 635-653.
3. A.T. Balaban, In Chemical Application of Graph Theory; Balaban, A. T., Ed.; Academic Press: London, 1976; Chapter 5, p 63.
4. A Balaban, J. Chem. Inf. Comput. Sci. (1985), **25**, 334-343.

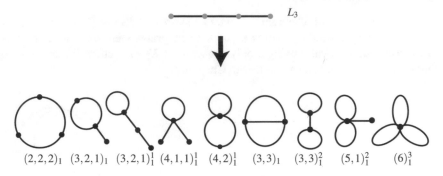

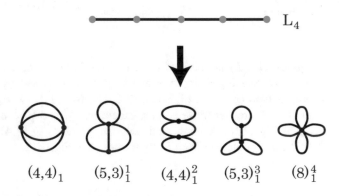

$$(2,2,2)_1 \quad (3,2,1)_1 \quad (3,2,1)_1^1 \quad (4,1,1)_1^1 \quad (4,2)_1^1 \quad (3,3)_1 \quad (3,3)_1^2 \quad (5,1)_1^2 \quad (6)_1^3$$

Fig. 2.16 Graphs which can be obtained from L_3 by folding.

$$(4,4)_1 \qquad (5,3)_1^1 \qquad (4,4)_1^2 \qquad (5,3)_1^3 \qquad (8)_1^4$$

Fig. 2.17 $G = (4,4)_1, (5,3)_1^1, (4,4)_1^2, (5,3)_1^3, (8)_1^4$ are obtained from L_4 by folding.

5. R.B. King and D.H. Rouvray, Eds., Graph Theory and Topology in Chemistry, Elsevier, Amsterdam, The Netherlands, 1987.
6. D.H. Rouvray, Chem. Soc. Rev. (1974), **3**, 355-372.
7. L.M. Masinter, N.S. Sridharan, J. Lederberg, and D.H. Smith, J. Am. Chem. Soc. (1974), **96**, 7702-7714.
8. L. Bytautas and D.J. Klein, Croat. Chem. Acta, (2000), **73**, 331-357.
9. D.J. Klein and H. Zhu, In Chemical Topology to Three-Dimensional Geometry, A.T. Balaban Ed., Plenum, New York, 1997.
10. D.S. Gaunt, J.E.G. Lipson, J.L. Martin, M.F. Sykes, G.M. Torrie, S.G. Whittington, and M.K.J. Wilkinson, Phys. A: Math. Gen. (1984), **17**, 211-236.
11. J.E. Martin and B.E. Eichinger, J. Chem. Phys. (1978), **69**, 4588-4594.
12. N. Lozac'h, A.L. Goodson, and W.H. Powell, Angew. Chem., Int. Ed. Engl. (1979), **18**, 887-899.
13. E. Flapan, When topology meets chemistry, Cambridge University Press, 2000.
14. E. Flapan, Knots, Molecules, and the Universe, American Mathematical Society, 2016.
15. Y. Tezuka, Ed., Topological Polymer Chemistry: Progress of cyclic polymers in syntheses, properties and functions (World Scientific, Singapore, 2013).
16. Y. Tezuka, Acc. Chem. Res., **50**, 2661 (2017).
17. A. Beukemann and W.A. Klee, Z. Kristallogr. **201**, 37-51 (1992).
18. A. Beukemann, Dissertation, Georg-August-Universität Göttingen, 2015.

19. G. R. Newkome, C. N. Moorefield and F. Vögtle, in Dendritic Molecules-Concepts, Synthe-ses, Perspectives (Wiley-VCH, Weinheim, 1996), p.37.
20. L. Euler, Solutio problematis ad geometriam situs pertinentis, Comment. Acad. Sci. Imp. Petrop **8**, 128-40 (1736).

Chapter 3
Classification of polymer topologies based on alkane molecular graphs

In this chapter, we describe the chemistry-based, hierarchical classification procedure in which a series of nonlinear, cyclic, and branched polymer architectures are classified from the molecular graph presentation of alkanes and cycloalkanes. We also discuss a systematic notation protocol for nonlinear polymer topologies, modified from the previous version ([3]), that is based on classification via the terminus T and junction J values in graph constructions of topological polymers.

3.1 Polymer graph constructions and alkane molecular graphs

In organic chemistry, the skeletal structures of alkanes (C_nH_{2n+2}) and a series of mono- and poly-cycloalkanes (e.g., C_nH_{2n}, C_nH_{2n-2}) are commonly represented by line graph forms [1, 2]. A systematic classification procedure for nonlinear polymer constructions and cyclic and multicyclic polymer structures of sufficiently long (and thus flexible) segments has been formulated based on graph presentations [3]. Topological relationships between different polymer structures are elucidated by the hierarchical classification, and their synthetic pathways can then be formulated. A comprehensive notation process will be developed, as in the cases of dendrimers, knots, catenanes, and rotaxanes [4, 5, 6].

As detailed below, the total number of termini (chain ends) and junctions (branch points) are invariant (constant) geometric parameters, as are the total number of branches and connectivities at each junction. The distance between two adjacent junctions and that between the junction and a terminus are variable geometrical parameters, in contrast to Euclidian geometry, and are in accord with flexible, randomly coiled, and constrained polymer segments. Contrary to molecular graphs based on tetravalent carbon atoms, polymer graph constructions having five or more branches at one junction are permitted.

© The Author(s), under exclusive license to Springer Japan KK 2019
K. Shimokawa et al., *Topology of Polymers*, SpringerBriefs in the Mathematics of Materials 4, https://doi.org/10.1007/978-4-431-56888-9_3

3.2 A nomenclature for alkane molecular graphs

In this section, we introduce a nomenclature for alkane molecular graphs.

Definition 3.1 (A nomenclature for molecular graphs of alkanes). We consider a molecular graph G of alkanes. Suppose the rank of G is r, the number of termini is T, and the number of junctions is J. We assume that G is a simple graph and has a minimal number of vertices. Let Y denote the number of vertices of G. We use $^{r}A_Y(T,J)$ to denote a molecular graph G. If there are isomers sharing the same nomenclature $^{r}A_Y(T,J)$, we add the degree sequence of the vertices.

We discuss the relationship of two nomenclatures in Section 3.6.

3.3 Classification of branched polymer topologies

The graph presentation of alkanes having the molecular formula C_nH_{2n+2}, where n=3-7, and selected higher alkanes having C_nH_{2n+2}, where n> 7, and their corresponding topological polymer constructions are hierarchically ranked in Table 3.1. A line construction is produced both from ethane (C_2H_6) and propane (C_3H_8); just the latter is listed for brevity.

Thus, the *n*- and *iso*-forms of butane produce linear and three-armed star branched polymer constructions, respectively. A *neo*-pentane produces a four-armed star graph construction in addition to the two obtained from butane. Similarly, H-shaped and five-armed star graph constructions are introduced from hexanes, as well as super H-shaped and six-armed stars from heptanes.

The systematic notation, modified from the previous version ([3]), has been given for a series of branched polymers listed in Table 3.1. They are labeled as ^{0}A main-class, because they are produced from alkanes free of cyclic units. A linear construction is produced from propane (C_3H_8), and this topology is ubiquitous in those produced from all higher alkanes. This sub-class construction is thus termed $^{0}A_3$, or, alternatively $^{0}A_3(2,0)$, by indicating the total number of termini and junctions, respectively, in parentheses. Similarly, sub-classes $^{0}A_4$ (or $^{0}A_4(3,1)$) and $^{0}A_5$ (or $^{0}A_5(4,1)$) are uniquely defined as in Table 3.1. Furthermore, multiple constructions are included in sub-classes $^{0}A_6$, $^{0}A_7$ and $^{0}A_8$, from hexanes, heptanes, and octanes, respectively, while each component in these classes is uniquely identified by indicating the total number of termini and junctions, as shown in Table 3.1. Following this process, an m-armed star polymer topology is labeled $^{0}A_{m+1}(m,1)$.

A pair of distinctive constructions of $^{0}A_9(6,3)$ are included in the sub-class $^{0}A_9$ from nonanes, in which the junction connections are distinct, while the total numbers of termini and junctions are identical. Thus, this pair is notated by $^{0}A_9(6,3)[4$-3-$3]$ and $^{0}A_9(6,3)[3$-4-$3]$, respectively, where a backbone chain having the most junctions is first identified, and then the number of branches at each junction is given in brackets in descending order from the most substituted junction. Another example is identified in the sub-class $^{0}A_{10}(6,4)$ from decanes, where one has a dendrimer-like

Topology	C_nH_{2n+2}					Topology	C_nH_{2n+2} $n=8$		Topology	C_nH_{2n+2} $n=10$
	n=3	4	5	6	7					
$^0A_3(2,0)$						$^0A_8(5,3)$			$^0A_{10}(6,4)[3\text{-}3(3)\text{-}3]$	
$^0A_4(3,1)$						$^0A_8(6,2)$			$^0A_{10}(6,4)[3\text{-}3\text{-}3\text{-}3]$	
$^0A_5(4,1)$						Topology	n=9		Topology	n=m
$^0A_6(4,2)$						$^0A_9(6,3)[4\text{-}3\text{-}3]$				
$^0A_6(5,1)$				$(*)$	$(\cancel{*})$	$^0A_9(6,3)[3\text{-}4\text{-}3]$			$^0A_{m+1}[m,1]$ m-Arm Star Polymer	
$^0A_7(5,2)$										
$^0A_7(6,1)$					$(*)$					

Table 3.1 Linear and branched polymer topologies produced from molecular graphs of alkanes (C_nH_{2n+2}: n = 3-7) and selected isomers (C_nH_{2n+2}: n = 8-10 and m).

star polymer structure and the other has a comb-like branched structure. They are notated as $^0A_{10}(6,4)[3\text{-}3(3)\text{-}3]$ and $^0A_{10}(6,4)[3\text{-}3\text{-}3\text{-}3]$, respectively.

3.4 Classification of monocyclic polymer topologies

A series of monocyclic polymer topologies are produced from molecular graphs of mono-cycloalkanes having the formula C_nH_{2n} up to n = 7, and are listed in Table 3.2. A simple cyclic (ring) graph construction is produced from cyclopropane (C_3H_6), and this graph construction is ubiquitous in all higher mono-cycloalkanes of C_nH_{2n}. The graph construction (tadpole) of a ring with a branch form is produced from methylcyclopropane, together with a simple ring from cyclobutane, with the chemical formula of C_4H_8. From the molecular graphs of cyclopentane isomers, C_5H_{10}, two polymer graph constructions are produced. One is a twin-tail "tadpole" having two outward branches at one common junction in the ring unit, and a two-tail tadpole having two outward branches located at two separate junctions in the ring unit. Four- and seven-polymer graph constructions are introduced by molecular graphs of cyclohexane isomers C_6H_{12} and cycloheptane isomers C_7H_{14}, respec-

Topology	C_nH_{2n}				C_nH_{2n}	
	n = 3	4	5	6	n = 7	
$^1A_3(0,0)$					$^1A_7(3,2)[1(4)]$	
$^1A_4(1,1)$					$^1A_7(3,2)[2(3,0)]$	
$^1A_5(2,1)$					$^1A_7(3,3)$	
$^1A_5(2,2)$					$^1A_7(4,1)$	
$^1A_6(2,2)$					$^1A_7(4,2)[3-1]$	
$^1A_6(3,1)$					$^1A_7(4,2)[2-2]$	
$^1A_6(3,2)$					$^1A_7(4,3)$	
$^1A_6(3,3)$						

Table 3.2 Monocyclic polymer topologies produced from molecular graphs of cycloalkanes (C_nH_{2n}: n=3-6) and selected cycloheptane (C_nH_{2n}: n=7) isomers.

tively. Those having five or more hypothetical branches at one junction are included (in parentheses) in Table 3.2.

The systematic notation is included for the series of monocyclic polymer topologies listed in Table 3.2. They are commonly labeled as a ^1A main-class because they are produced from monocycloalkanes. Thus, a simple ring construction produced from cyclopropane and higher cycloalkanes is designated as sub-class $^1A_3$ or, alternatively, $^1A_3(0,0)$ by indicating the total number of termini and junctions in parentheses. Similarly, a tadpole graph construction is labeled $^1A_4$ or $^1A_4(1,1)$ because it initially refers to methylcyclopropane, and has the formula C_4H_8. Each construction produced in the sub-classes $^1A_5$ and $^1A_6$ are uniquely labeled simply by specifying the total number of termini and junctions in the corresponding graph constructions.

A pair of distinctive polymer graph constructions are included in the sub-classes $^1A_7(3,2)$ and $^1A_7(4,2)$, produced from mono-cycloalkanes having the formula C_7H_{14}; they have distinctive branch arrangements on a ring unit. Thus, the number of outward branches at each junction on the ring unit is identified and indicated in the brackets placed after the closing parenthesis. A detailed junction architecture of the outward branches is indicated in parentheses enclosed within the brackets, to identify the connectivity of junctions along the ring unit with the number of outward branches at each junction by connecting them with a hyphen. The two constructions of $^1A_7(3,2)$ and $^1A_7(4,2)$ are thereby notated specifically as $^1A_7(3,2)[1(4)]$ and $^1A_7(3,2)[2(3,0)]$, and as $^1A_7(4,2)[3-1]$ and $^1A_7(4,2)[2-2]$, respectively, as listed in Table 3.2.

3.5 Classification of multicyclic polymer topologies

3.5.1 Dicyclic polymer topologies

A class of dicyclic polymer topologies are produced from molecular graphs of bicycloalkanes, and those with C_nH_{2n-2}, up to n = 6, are listed in Table 3.3. The three specific constructions are regarded as basic forms because they possess no free chain ends and contain no outward branches. More specifically, they are an internally linked (fused) θ-form, a directly linked (spironic) 8-form, and an externally linked (bridged) manacle-form.

A *fused*-dicyclic (θ-form) construction is produced first from bicyclo[1,1,0]butane, C_4H_6. From the five molecular graphs of bicyclopentane isomers with the formula C_5H_8, three graph constructions are introduced, including a *spiro*-dicyclic (8-shaped) construction from spiro[2,2]pentane. Moreover, eight graph constructions are produced by reference to bicyclohexane isomers (C_6H_{10}), including a *bridged*-dicyclic (manacle-form) produced from the molecular graph of bi(cyclopropane).

The systematic notation is included for a series of dicyclic polymer topologies listed in Table 3.3. They are labeled as a ^2A main-class because they are produced from di-cycloalkanes. Thus, a θ-form is defined as sub-class $^2A_4$, or $^2A_4(0,2)$ by indicating the total number of termini and junctions, respectively, in parentheses. Three and eight constructions are introduced in the sub-classes $^2A_5$ and $^2A_6$, respectively, as listed in Table 3.3. In sub-classes $^2A_6(2,2)$ and $^2A_6(2,3)$, each construction is identified by the number of outward branches as well as the number of internally linked branches on the ring unit. Thus, in the brackets after the closing parenthesis, the numbers of outward and internally linked branches on the ring unit are indicated, and these numbers (thus 0 for the latter, and $1, 2, \cdots$ for the former) are linked by hyphens. The positions of the two specific junctions internally

Topology	C_nH_{2n-2}			Topology	C_nH_{2n-2}
	n = 4	5	6		n = 6
$^2A_4(0,2)$				$^2A_6(0,2)$	
$^2A_5(0,1)$				$^2A_6(1,1)$	
$^2A_5(1,2)$				$^2A_6(1,2)$	
$^2A_5(1,3)$				$^2A_6(2,2)[2^a\text{-}0^a]$	
				$^2A_6(2,2)[1^a\text{-}1^a]$	
				$^2A_6(2,3)[2\text{-}0^a\text{-}0^a]$	
				$^2A_6(2,3)[1^a\text{-}1\text{-}0^a]$	
				$^2A_6(2,4)$	

Table 3.3 Dicyclic polymer topologies produced from molecular graphs of bicycloalkanes (C_nH_{2n-2}: n = 4-6).

linked to each other are indicated by superscripts (e.g., a, b) at the relevant junction numbers. For examples, the two $^2A_6(2,2)$'s are uniquely notated by: $^2A_6(2,2)[2^a\text{-}0^a]$ and $^2A_6(2,2)[1^a\text{-}1^a]$ and the two $^2A_6(2,3)$'s are defined as $^2A_6(2,3)[2\text{-}0^a\text{-}0^a]$ and $^2A_6(2,3)[1^a\text{-}1\text{-}0^a]$, respectively.

3.5.2 Tricyclic polymer topologies

Topology	C_nH_{2n-4}				Topology	C_7H_{10}
	n = 4	5	6	7		
$^3A_4(0,4)[0^a\text{-}0^b\text{-}0^a\text{-}0^b]$					$^3A_7(0,3)$	
					$^3A_7(0,4)$	
$^3A_5(0,2)[0^{a,b}\text{-}0^{a,b}]$					Topology	C_8H_{12}
$^3A_5(0,3)[0^{a,b}\text{-}0^a\text{-}0^b]$					$^3A_8(0,2)$	
					$^3A_8(0,3)$	
$^3A_6(0,4)[0^a\text{-}0^a\text{-}0^b\text{-}0^b]$					Topology	C_9H_{14}
$^3A_6(0,2)$					$^3A_9(0,3)$	
$^3A_6(0,3)$					$^3A_9(0,4)$	
$^3A_7(0,1)$					Topology	$C_{10}H_{16}$
$^3A_7(0,2)$					$^3A_{10}(0,4)$	

Table 3.4 Tricyclic polymer topologies produced from molecular graphs of tri-cycloalkanes without outward branches (C_nH_{2n-4}: n = 4-10).

A class of tricyclic polymer topologies are produced from molecular graphs of tri-cycloalkanes in reference to tri-cycloalkane isomers having the formula C_nH_{2n-4}. Fifteen constructions are regarded as basic forms because they possess no free chain ends or outward branches. They are produced from the relevant molecular graphs of tri-cycloalkane isomers of C_nH_{2n-4}, with n=4-10, as listed in Table 3.4. A large number of tricyclic polymer graph constructions having outward branches are omitted for brevity. More specifically, four tricyclic constructions are in the *fused*-form, two are in the *spiro*-form, and three are in the *bridged*-form, in addition to the six *hybrid*-forms by the combination of either *fused/spiro*, *fused/bridged*, or *spiro/bridged* forms.

The systematic notation is included for a series of tricyclic polymer topologies listed in Table 3.4. They are labeled ^3A main-class because they are produced from tri-cycloalkanes. Thus, a doubly-*fused* (internally double-linked) ring construction is introduced from tetrahedrane and is designated as a sub-class $^3A_4$ or $^3A_4(0,4)$ by indicating the total number of termini and junctions, respectively, in parentheses. All four doubly-fused tricyclic graph constructions, $^3A_4(0,4)$, $^3A_5(0,2)$, $^3A_5(0,3)$, and $^3A_6(0,4)$, are α, δ, γ and β graphs, respectively [4].

In a series of doubly-*fused* constructions of $^3A_4(0,4)$, $^3A_5(0,2)$, and $^3A_5(0,3)$, the arrangements in internally linked branches on the ring unit are uniquely defined as $[0^a-0^b-0^a-0^b]$, $[0^{a,b}-0^{a,b}]$, and $[0^{a,b}-0^a-0^b]$, respectively. Other tricyclic constructions having either *spiro*-, *bridged*-, or *hybrid*-forms listed in Table 3.4 are uniquely notated by simply indicating the total number of termini and junctions in each graph construction.

3.5.3 Tetra- and pentacyclic polymer topologies

Finally, five examples of topologically significant *fused*-tetracyclic constructions and a *fused*-pentacyclic construction are listed in Table 3.5. These are produced by reference to the relevant tetra- and penta-cycloalkane isomers having the formula C_nH_{2n-6}, with n=6, and of C_nH_{2n-8}, with n=8, respectively. Thus, they are labeled ^4A and ^5A main-class, respectively, in which a $K_{3,3}$ graph construction is included. It is a prototypical non-planar graph that is incapable of embedding in a plane without crossing the edges. The $K_{3,3}$ graph construction is uniquely notated as $^4A_6(0,6)[0^a-0^b-0^c-0^a-0^b-0^c]$, as well as other *triply*-fused tetracyclic constructions as $^4A_6(0,3)[0^{a,c}-0^{a,b}-0^{b,c}]$ for the "unfolded tetrahedron" construction, as $^4A_6(0,6)[0^a-0^b-0^a-0^c-0^b-0^c]$ for the Prisman construction, as $^4A_8(0,6)[0^a-0^a-0^b-0^c-0^c-0^b]$ for the ladder construction, and as $^4A_6(0,4)[0^{a,b,c}-0^a-0^b-0^c]$ for the "Königsberg bridge" construction, respectively, in which the mode of internally linked branches on the ring unit is uniquely characterized. By a similar manner, a quadruply-*fused* pentacyclic graph construction of a "shippo" form is produced and notated as: $^5A_8(0,4)[0^{b,d}-0^{a,c}-0^{b,d}-0^{a,c}]$, as seen in Table 3.5. Chemical synthesis of these complex polymer topologies are covered in Chapter 7.

Topology	C_6H_6	Topology	C_6H_6
$^4A_6(0,3)[0^{a,c}\text{-}0^{a,b}\text{-}0^{b,c}]$	Unfolded tetrahedron	$^4A_8(0,6)[0^a\text{-}0^a\text{-}0^b\text{-}0^c\text{-}0^c\text{-}0^b]$	Ladder
$^4A_6(0,6)[0^a\text{-}0^b\text{-}0^a\text{-}0^c\text{-}0^b\text{-}0^c]$	Prisman	$^4A_6(0,4)[0^{a,b,c}\text{-}0^a\text{-}0^b\text{-}0^c]$	Koenigsberg bridge
$^4A_6(0,6)[0^a\text{-}0^b\text{-}0^c\text{-}0^a\text{-}0^b\text{-}0^c]$	$K_{3,3}$	$^5A_8(0,4)[0^{b,d}\text{-}0^{a,c}\text{-}0^{b,d}\text{-}0^{a,c}]$	Shippo

Table 3.5 Selected tetracyclic and pentacyclic polymer topologies produced from tetra-and penta-cycloalkanes (C_nH_{2n-6}: n = 6, and C_nH_{2n-8}: n = 8,).

3.6 Comparison of nomenclatures

We compare two nomenclatures for polymers given in Chapters 2 and 3. We assume graphs without terminus, i.e., T = 0.

For dicyclic polymers, $^2A_m(0,n)$ corresponds to three dicyclic graphs in Theorem 2.1. In case $n = 1$, then $^2A_m(0,1)$ is $(4)_1^2$. In case $n = 2$, then $^2A_m(0,2)$ is either $(3^2)_1$ or $(3^2)_1^2$.

For tricyclic polymers, $^3A_m(0,n)$ corresponds to 15 tricyclic graphs in Theorem 2.2. In case $n = 1$, then $^3A_m(0,1)$ is $(6)_1^3$. In case $n = 2$, then $^3A_m(0,2)$ is either $(4^2)_1$, $(5,3)_1^1$, $(6^3)_1$, $(4^2)_1^2$, or $(5,3)_1^3$. In case $n = 3$, then $^3A_m(0,3)$ is either $(4,3^2)_1$, $(4,3^2)_1^1$, $(4,3^2)_2^1$, $(4,3^2)_1^2$, or $(4,3^2)_1^3$. In case $n = 4$, then $^3A_m(0,4)$ is either $(3^4)_1$, $(3^4)_2$, $(3^4)_1^1$, $(3^4)_1^2$ or $(3^4)_1^3$.

For tetracyclic polymers in Table 3.5, $^4A_6(0,3)[0^{a,c}\text{-}0^{a,b}\text{-}0^{b,c}]$ is $(4^3)_1$, $^4A_6(0,6)[0^a\text{-}0^b\text{-}0^a\text{-}0^c\text{-}0^b\text{-}0^c]$ is $(3^6)_2$, $^4A_6(0,6)[0^a\text{-}0^b\text{-}0^c\text{-}0^a\text{-}0^b\text{-}0^c]$ is $(3^6)_1$, $^4A_8(0,6)[0^a\text{-}0^a\text{-}0^b\text{-}0^c\text{-}0^c\text{-}0^b]$ is $(3^6)^4$, and $^4A_6(0,4)[0^{a,b,c}\text{-}0^a\text{-}0^b\text{-}0^c]$ is $(5,3^3)_1$. The degree sequence of the pentacyclic polymer $^5A_8(0,4)[0^{b,d}\text{-}0^{a,c}\text{-}0^{b,d}\text{-}0^{a,c}]$ is (4^4).

References

1. A. T. Balaban, in Chemical Application of Graph Theory, Ed. A. T. Balaban, (Academic Press, London, 1976), p.63.
2. R. B. King and D. H. Rouvray, Eds., Graph Theory and Topology in Chemistry (Elsevier, Amsterdam, 1987).G. R. Newkome, C. N. Moorefield, and F. Vögtle, in Dendritic Molecules Concepts, Syntheses, Perspectives (Wiley-VCH, Weinheim, 1996), p.37.
3. Y. Tezuka and H. Oike, J. Am. Chem. Soc., **123**, 11570 (2001).

4. G. R. Newkome, C. N. Moorefield, and F. Vögtle, in Dendritic Molecules-Concepts, Synthe-
    ses, Perspectives (Wiley-VCH, Weinheim, 1996), p.37.
5. T. Schucker, New J. Chem., **17**, 655 (1993).
6. J.-C. Chambron, C. Dietrich-Buchecker, and J.-P. Sauvage, in Cyclic Polymers, 2nd ed., Ed.
    J. A. Semlyen, (Kluwer, Dordorecht, 2000), p.155.

Chapter 4
Types of graphs

As seen in Chapter 2, topology variety increases with increasing multicyclic graphs: e.g., dicyclic, tricyclic, tetracyclic. Multicyclic graphs are classified into *spiro*, *bridge*, *fused*, and *hybrid* types [1]. In this chapter, we present graph theory definitions of the various multicyclic graphs and characterize each type via construction and decomposition. The characterizations may suggest polymer synthesis procedures. See Chapter 7 for the synthesis of multicyclic polymers.

4.1 Operations and decomposition of graphs

In this section, first we introduce operations and decomposition of graphs to understand topological construction. Next, focusing on dicyclic subgraphs, we define the types (Definition 4.1). Finally, we construct multicyclic graph structures from a loop (Theorem 4.1).

4.1.1 Operations of graphs

We introduce operations that produce a construction of graphs from a simpler one. Let $G_1 = (V_1, E_1), G_2 = (V_2, E_2)$ and $G = (V, E)$ be graphs.

Graph union: A graph union $G_1 \cup G_2$ of G_1 and G_2 is a graph $(V_1 \cup V_2, E_1 \cup E_2)$. In particular, the graph union is a *disjoint graph union* if V_1 and V_2 are disjoint. If $V_1 \cap V_2$ is a singleton set $\{v\}$, then the graph union $G_1 \cup G_2$ is called a *wedge sum* of G_1 and G_2 on v, and it is denoted by $G_1 \vee G_2$ or $G_1 \vee_v G_2$. The vertex v of the wedge sum $G_1 \vee_v G_2$ is a *cut-vertex* of $G_1 \vee_v G_2$. The wedge sum $G_1 \vee_v G_2$ is *cyclic* if G_1 and G_2 have cycles containing v.

Subdivision: A *subdivision* $G * e$ of edge e in G is a graph $(V \cup \{v'\}, (E - \{e\}) \cup \{e'_1, e'_2\})$, where $e \in E$ connects $v_1 \in V$ and $v_2 \in V$, $e'_i \notin E$ $(i = 1, 2)$ are edges

K. Shimokawa et al., *Topology of Polymers*, SpringerBriefs in the Mathematics of Materials 4, https://doi.org/10.1007/978-4-431-56888-9_4

that connect v_i and v'. In general, a subdivision of graph G is a graph resulting from the subdivision of edges in G. The subdivision of G is homeomorphic to G because each edge of G is preserved or just replaced by a path. If two graphs G_1 and G_2 are homeomorphic, a subdivision of G_1 is isomorphic to a subdivision of G_2.

Smoothing: A *smoothing* of a vertex is the reverse operation of an edge subdivision. Namely, a smoothing of vertex $v \in V$ with $d(v) = 2$ is a graph $(V - \{v\}, (E - \{e_1, e_2\}) \cup \{e'\})$, where $e_i \in E$ connects $v_i \in V$ and v $(i = 1, 2)$, and $e' \notin E$ is an edge which connects v_1 and v_2.

Edge addition: An *edge addition* $G + e'$ of G is a graph $(V, E \cup \{e'\})$, where $e' \notin E$ and $\partial e' \subset V$. There are three cases:

a. In the case where $\partial e' = \{v\}$ (e' is a loop), $G + e'$ is a *loop addition* (*L-addition*) of G. In this case, $G + e'$ is a wedge sum $G \vee_v L$ of G and a loop $L = (\{v\}, \{e'\})$. The loop addition $G + e'$ is a *cyclic loop addition* (*CL-addition*) if G has a cycle including the vertex v.

b. In the case where $\partial e' = \{v_1, v_2\}$ and there are no paths connecting v_1 and v_2 in G, G is a disjoint graph union of two graphs $G_1 = (V_1, E_1)$ and $G_2 = (V_2, E_2)$, so that $v_i \in V_i$ $(i = 1, 2)$. In this case, $G + e'$ is a *bridge sum* of G_1 and G_2, and is denoted by $G_1 -_{e'} G_2$. The edge e' is called a *bridge* of $G + e'$. By putting $\overline{G_i} = (V_i \cup \{v_1, v_2\}, E_i \cup \{e'\})$, we have $G + e' = G_1 \vee_{v_1} \overline{G_2} = \overline{G_1} \vee_{v_2} G_2$. If $G_2 = (\{v_2\}, \emptyset)$, $G + e'$ is a *pendant addition* (*P-addition*) of graph G_1 because the vertex v_2 and the edge e' are a pendant vertex and a pendant edge of $G + e'$, respectively. Note that a pendant edge is a bridge. A bridge sum $G_1 -_{e'} L$ of G_1 and a loop $L = (\{v_2\}, \{e''\})$ is a *loop-bridge addition* (*LB-addition*) of G_1

c. In the case where $\partial e' = \{v_1, v_2\}$ and there is a path connecting v_1 and v_2 in G, $G + e'$ is a *bypass edge addition* (*BE-addition*) of G. In particular, if G has a cycle including both v_1 and v_2, we call the BE-addition a *cyclic edge addition* (*CE-addition*) of G.

Edge elimination: An *edge elimination* is the reverse operation of an edge addition. Namely, an edge elimination of G is a graph $G - e = (V, E - \{e\})$, where $e \in E$.

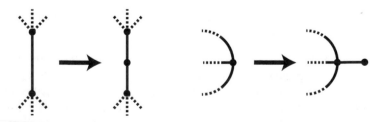

Fig. 4.1 A subdivision of an edge and a P-addition

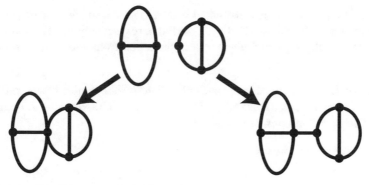

Fig. 4.2 A (cyclic) wedge sum and a bridge sum

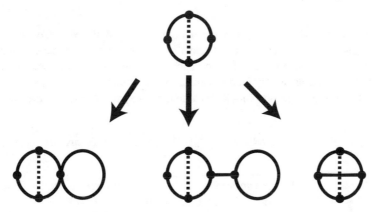

Fig. 4.3 A CL-addition, an LB-addition, a CE-addition

On each operation of graphs, a change of the rank is easily seen by Proposition 2.1:

Lemma 4.1. *For a subdivision $G * e$ and a P-addition $G + e'$ of G,*

$$r(G * e) = r(G + e') = r(G)$$

For a wedge sum $G_1 \vee_v G_2$ and a bridge sum $G_1 -_{e'} G_2$ of G_1 and G_2,

$$r(G_1 \vee_v G_2) = r(G_1 -_{e'} G_2) = r(G_1) + r(G_2)$$

In particular, a CL-addition, an LB-addition, and a CE-addition increase the rank by one.

Any tree (resp. monocyclic graph) can be constructed from a vertex (resp. loop) by a sequence of P-additions and subdivisions. Similarly, any dicyclic graphs can be obtained from the three dicyclic graphs $(4)_1^2$, $(3^2)_1^2$ and $(3^2)_1$ in Theorem 2.1. The three dicyclic graphs can be constructed from a loop or a subdivision of a loop by a CL-addition, an LB-addition, and a CE-addition, respectively, see Fig. 4.3. For any multicyclic graph, in general, there is a construction from a loop using operations above:

Theorem 4.1. *[The construction method of multicyclic graphs] Any multicyclic graph can be constructed from a loop by a sequence of CL-additions, LB-additions, CE-additions, P-additions, and subdivisions.*

We will prove Theorem 4.1 later.

4.1.2 Blocks and types

A maximal connected subgraph without a cut-vertex of the subgraph is a *block*. Each bridge is a block because each end is a cut-vertex or a pendant vertex. Two maximal blocks of graph G overlap in at most one vertex, which is a cut-vertex of G. Hence, every edge is contained in precisely one block, and any connected graph is uniquely decomposed into blocks by maximally applying the reverse operations of wedge sums.

Lemma 4.2. *Blocks of a wedge sum $G_1 \vee_v G_2$ of G_1 and G_2 are blocks of G_1 and blocks of G_2.*

Proof. A maximal connected subgraph of $G_1 \vee_v G_2$ without a cut-vertex of the subgraph is that of G_1 or that of G_2 because v is a cut-vertex of $G_1 \vee_v G_2$.

Every cycle of G is contained in exactly one block because the cycle has no cut-vertex of itself. Conversely, each block other than a bridge contains a cycle. More precisely, each edge other than a bridge is extended into a cycle by adding a path:

Lemma 4.3. *Let H be a subgraph of block B with at least two vertices and $H \neq B$. Then, there is an H-path in B.*

Here, an H-path is a path whose ends are in H, and the other vertices and edges are not in H.

Proof. Let $H_1, \cdots, H_k$ be maximally connected subgraphs without an edge of H. By their maximal nature, they are mutually disjoint. Since B is connected, each H_i is sharing at least one vertex with H. In fact, each H_i, say H_1, is sharing at least two vertices with H; otherwise, B is a wedge sum of H_1 and $H \cup H_2 \cup \cdots \cup H_k$, a contradiction. Thus H_1 contains an H-path since H_1 is connected.

A monocyclic graph has precisely one block, which is a cycle, and the other blocks are bridges. Blocks of dicyclic graphs $(4)_1^2$, $(3^2)_1^2$, and $(3^2)_1$ are two loops, two loops and one bridge, and $(3^2)_1$ itself, respectively. We introduce elementary dicyclic subgraphs for multicyclic graphs:

Proper 8-subgraph An *8-subgraph* is a subgraph of G that is homeomorphic to the graph $(4)_1^2$ (*i.e.*, and isomorphic to a subdivision of $(4)_1^2$). An 8-subgraph is *proper* if the cut-vertex of the 8-subgraph is a cut-vertex of G.

Proper manacle-subgraph A *manacle-subgraph* of G is a subgraph which is homeomorphic to the graph $(3^2)_1^2$ (*i.e.*, isomorphic to a subdivision of $(3^2)_1^2$). A manacle-subgraph is *proper* if each bridge of the manacle-subgraph is a bridge of G.

θ-subgraph A *θ-subgraph* of G is a subgraph which is homeomorphic to the graph $(3^2)_1$ (*i.e.* isomorphic to a subdivision of $(3^2)_1$).

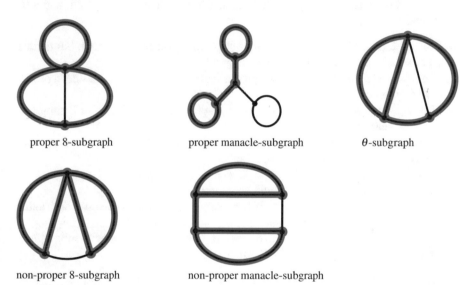

proper 8-subgraph   proper manacle-subgraph θ-subgraph

non-proper 8-subgraph   non-proper manacle-subgraph

Fig. 4.4 Proper/non-proper 8-subgraphs, proper/non-proper manacle-subgraphs, and a θ-subgraph

Proper 8-subgraphs, proper manacle-subgraphs, and θ-subgraphs are together *elementary (dicyclic) subgraphs*. A θ-subgraph of G is contained in a block because it has no cut-vertex of itself. Conversely,

Lemma 4.4. *A block B which is neither a bridge nor a cycle contains a θ-subgraph.*

Proof. Because B is not a bridge, it contains a cycle H. Because $H \neq B$, B contains an H-path P by Lemma 4.3. Then H and P form a θ-subgraph that is contained in B.

Any proper 8-subgraph extends over two blocks of G containing cycles, because the cut-vertex of itself is a cut-vertex of G. Similarly, any proper manacle-subgraph extends over two blocks of G containing cycles and bridge(s) of G, because every end of the bridge(s) of the proper manacle-subgraph are cut-vertices of G.

Lemma 4.5. *Suppose a multicyclic graph G has a block C that is a cycle. Then G has a proper 8-subgraph containing C or a proper manacle-subgraph containing C.*

Proof. Since G is a multicyclic graph, G contains a cycle other than C. All blocks with cycles, including C, are connected by cut-vertices and paths consisting of bridges. If a cycle C' shares a cut-vertex with C, then C and C' form a proper 8-subgraph of G. Otherwise, there is a cycle C' and a path P consisting of bridges such that P connects a vertex of C and a vertex of C'. Then C, P, and C' form a proper manacle-subgraph of G.

By Lemmas 4.4 and 4.5, any multicycle graph contains at least one of a proper 8-subgraph, a proper manacle-subgraph, and a θ-subgraph. Then we can classify multicyclic graphs into the following seven types, according to whether or not the graphs contain each of a proper 8-subgraph, a proper manacle-subgraph, and a θ-subgraph.

Definition 4.1. 1. G is of *spiro type* if G contains a proper 8-subgraph, but neither a proper manacle-subgraph nor a θ-subgraph.
2. G is of *bridged type* if G contains a proper manacle-subgraph, but neither a proper 8-subgraph nor a θ-subgraph.
3. G is of *fused type* if G contains a θ-subgraph, but neither a proper 8-subgraph nor a proper manacle-subgraph.
4. G is of *spiro/bridged hybrid type* if G contains a proper 8-subgraph and a proper manacle-subgraph, but not a θ-subgraph.
5. G is of *spiro/fused hybrid type* if G contains a proper 8-subgraph and a θ-subgraph, but not a proper manacle-subgraph.
6. G is of *bridged/fused hybrid type* if G contains a proper manacle-subgraph and a θ-subgraph, but not a proper 8-subgraph.
7. G is of *spiro/bridged/fused hybrid type* if G contains a proper 8-subgraph, a proper manacle-subgraph, and a θ-subgraph.

Example 4.1. The graphs $(4)_1^2, (4^2)_1^2, (6)_1^3$ are of spiro type. The graphs $(3^2)_1^2, (3^4)_1^2, (4, 3^2)_1^3, (3^4)_1^3$ are of bridged type. The graphs $(3^2)_1, (4^2)_1, (4, 3^2)_1, (3^4)_1, (3^4)_2$ are of fused type. See Fig. 4.5.

Using the block decomposition, we can give a characterization of spiro, bridged, and fused types for graphs without a pendant vertex, i.e., graphs with degree of at least 2 at each vertex.

Theorem 4.2. *Suppose G is a multicyclic graph without a pendant vertex.*

1. *G is of spiro type if and only if every block is a cycle.*
2. *G is of bridged type if and only if every block is a bridge or a cycle, and every cut-vertex is contained in at most one cycle.*
3. *G is of fused type if and only if G has no cut-vertex.*

We will prove Theorem 4.2 and and focus on blocks which include exactly one cut-vertex to prove Theorem 4.1.

Lemma 4.6. *Suppose a multicyclic graph G has a cut-vertex. Then there exists a block B including exactly one cut-vertex v of G that satisfies one of the following.*

Spiro type

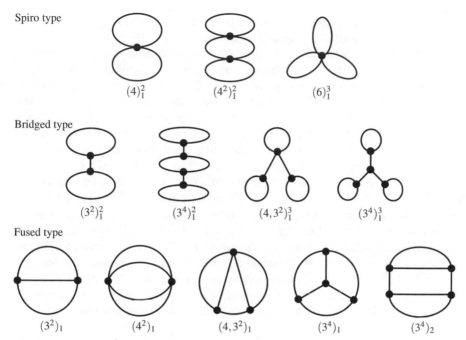

$(4)_1^2$ $(4^2)_1^2$ $(6)_1^3$

Bridged type

$(3^2)_1^2$ $(3^4)_1^2$ $(4,3^2)_1^3$ $(3^4)_1^3$

Fused type

$(3^2)_1$ $(4^2)_1$ $(4,3^2)_1$ $(3^4)_1$ $(3^4)_2$

Fig. 4.5 Graphs of spiro, bridged, fused type

(1) B is a pendant edge of G.
(2) B is a cycle contained in a proper 8-subgraph.
(3) B is a cycle and $d(v) = 3$ in G.
(4) B contains a θ-subgraph.

Proof. By Lemma 4.2, G is obtained from a block by a sequence of wedge sums adding blocks. Then the block which is added in the end of the sequence has exactly one cut-vertex of G. Suppose that each block including exactly one cut-vertex of G does not satisfy any of (1), (2) and (4) in Lemma 4.6. Then every block including exactly one cut-vertex of G is a cycle. If two of such blocks share their cut-vertices, then they form a proper 8-subgraph, which contradicts the assumption. Hence we have a connected graph G' by removing all such blocks. Let B' be a block including exactly one cut-vertex of G' if G' has a cut-vertex, otherwise let $B' = G'$. B' must have a cut-vertex v of G that is not a cut-vertex of G'. There is a removed block B of G that includes v. Since B is a cycle and is not contained in a proper 8-subgraph, B' has to be a bridge of G and $d(v) = 3$ in G. Then, block B with cut-vertex v satisfies (3) in Lemma 4.6.

Now we prove Theorem 4.1:

Proof. This proof is by mathematical induction on the rank r of a multicyclic graph G. By maximally applying smoothing and P-elimination, which are the reverse operations of subdivisions and P-additions, we may assume that G has no vertex with degree at most 2. In the case of $r = 2$, G is isomorphic to one of $(4)_1^2$, $(3^2)_1^2$ and $(3^2)_1$ that are obtained from a loop by a CL-addition, an LB-addition, and a CE-addition after a subdivision, respectively. Suppose $r > 2$ and any multicyclic graph with rank $r - 1$ can be constructed from a loop by a sequence of CL-additions, LB-additions, CE-additions, P-additions, and subdivisions. If G has no cut-vertex, then G itself is a block. Lemma 4.3 implies that G can be constructed from a cycle by adding paths inductively. The property that any two vertices included in a cycle is invariant under this construction. Thus, the last addition of path is a CE-addition of a connected graph G' with $r(G') = r - 1$. If G has a cut-vertex, then there is a block B, including exactly one cut-vertex v of G, that satisfies (2), (3), or (4) in Lemma 4.6. In case (2), because B is a loop and G' has a cycle containing v, G is obtained from a connected graph G' with $r(G') = r - 1$ by a CL-addition. In case (3), since B is a loop and exactly one bridge of G is adjacent to v, G is obtained from a connected graph G' with $r(G') = r - 1$ by a LB-addition. In case (4), by the same argument as above, B can be constructed from a cycle including v by adding paths inductively. The last addition of an H-path for a subgraph H of B becomes a CE-addition of a connected graph G' with $r(G') = r - 1$, because v is contained in H. Thus, by the inductive hypothesis, G can be constructed in all these cases from a loop by a sequence of CL-additions, LB-additions, CE-additions, P-additions and subdivisions.

4.2 Graph theory characterization of graph types

The construction sequence from a loop in Theorem 4.1 is not uniquely determined for a given multicyclic graph. However, the types of CL-, LB-, and CE-additions are uniquely determined from the type of initial graph. In other words, whether each of the CL-, LB-, and CE-additions is used in the construction is uniquely determined by the G type:

Theorem 4.3. *1. A P-addition has no effect on the presence of each proper 8-subgraph, proper manacle-subgraph, and θ-subgraph.*
2. A subdivision has no effect on the presence of each proper 8-subgraph, proper manacle-subgraph, and θ-subgraph.
3. A cyclic wedge sum, especially a CL-addition, produces a proper 8-subgraph.
4. A bridge sum of two monocyclic or multicyclic graphs, especially an LB-addition of a monocycle or multicyclic graph, produces a proper manacle-subgraph.
5. A CE-addition produces a θ-subgraph.
6. A cyclic wedge sum, especially a CL-addition, has no effect on the presence of a proper manacle-subgraph and a θ-subgraph.
7. A bridge sum, especially a LB-addition, has no effect on the presence of a proper 8-subgraph and a θ-subgraph.
8. A CE-addition has no effect on the presence of a proper 8-subgraph and a proper manacle-subgraph.

Proof. 1. Since any elementary subgraph contains neither a pendant edge nor a pendant vertex, a P-addition has no effect on the presence of a proper 8-subgraph, a proper manacle-subgraph, and a θ-subgraph.

2. Because a subdivision operation does not change the homeomorphism type of the graph and its subgraphs, it has no effect on the presence of a proper 8-subgraph, a proper manacle-subgraph and a θ-subgraph.

3. For a cyclic wedge sum $G_1 \vee_v G_2$, by definition, G_i contains a cycle C_i including v for each $i \in \{1,2\}$. Then C_1 and C_2 form a proper 8-subgraph $C_1 \vee_v C_2$ of $G_1 \vee_v G_2$.

4. Let $G_1 -_{e'} G_2$ be a bridge sum of two monocyclic or multicyclic graphs G_1 and G_2, where ends v_1 and v_2 of e' are contained in G_1 and G_2, respectively. For each $i \in \{1,2\}$, we define a subgraph C_i of G_i. If G_i contains a cycle having v_i, then let C_i be the cycle. Otherwise, G_i contains a cycle and a path P_i connecting a vertex of the cycle and v_i, where P_i consists of bridges. Let C_i be the union of the cycle and the path P_i in that case. Then C_1, e' and C_2 form a proper manacle-subgraph.

5. For a CE-addition $G + e'$ of G, by definition, G contains a cycle C which includes both ends of e'. Then C and e' form a θ-subgraph of $G + e'$.

6. Clearly, every elementary subgraph of G_i is an elementary subgraph of $G_1 \vee_v G_2$ for each $i \in \{1,2\}$. Conversely, every θ-subgraph of $G_1 \vee_v G_2$ is a θ-subgraph of G_1 or G_2. For the remaining case, suppose a cyclic wedge sum $G_1 \vee_v G_2$ contains a proper manacle-subgraph G'. We may assume that G' is neither a subgraph of G_1 nor a subgraph of G_2, otherwise the conclusion holds. The two cycles of G' are split into G_1 and G_2, and each bridge of G' is a bridge of exactly one of G_1 and G_2. Without loss of generality, we may assume that G_1 contains a bridge of G'. Let C and P be the cycle of G' in G_1 and the union of all bridges of G' in G_1. Since the wedge sum $G_1 \vee_v G_2$ is cyclic, G_1 contains a cycle C_1 having v. Then C, P, and C_1 form a proper manacle-subgraph.

7. Clearly, every elementary subgraph of G_i is an elementary subgraph of $G_1 -_{e'} G_2$ for each $i \in \{1,2\}$. Conversely, every θ-subgraph of $G_1 -_{e'} G_2$ is a θ-subgraph of G_1 or G_2 since any θ-subgraph is contained in a block. Similarly, every proper 8-subgraph of $G_1 -_{e'} G_2$ is a proper 8-subgraph of G_1 or G_2 since any proper 8-subgraph consists of two cycles in distinct blocks sharing a cut-vertex.

8. Clearly, every proper 8-subgraph and every proper manacle-subgraph of G are subgraphs of a CE-addition $G + e'$. Conversely, suppose a CE-addition $G + e'$ of G contains a proper 8-subgraph or a proper manacle-subgraph G', where G contains a cycle C having both ends of e'. We may assume that G' contains e', otherwise G' is a subgraph of G. The edge e' is contained in one of two cycles of G'. The two ends of e' split C into two paths, and at least one P of the two paths is not contained in G'. Then we obtain a subgraph of G from G' by replacing e' with P, that is a proper 8-subgraph or a proper manacle-subgraph if G' is a proper 8-subgraph or a proper manacle-subgraph, respectively.

The following corollary follows from Theorems 4.1 and 4.3.

Corollary 4.1. *1. A multicyclic graph G is a spiro type if and only if G is constructed from a loop by a sequence of CL-additions, P-additions and subdivisions.*
*2. A multicyclic graph G is a bridged type if and only if G is constructed from a
loop by a sequence of LB-additions, P-additions and subdivisions.*
*3. A multicyclic graph G is a fused type if and only if G is constructed from a loop
by a sequence of CE-additions, P-additions and subdivisions.*

Now we prove Theorem 4.2.

Proof. 1. Every proper manacle-subgraph contains a bridge, and every θ-subgraph
   is contained in a block. Thus, if every block is a cycle, then G has neither a
   proper manacle-subgraph nor a θ-subgraph; hence G is a spiro type. Conversely,
   suppose G is spiro type. By Corollary 4.1, G is constructed from a loop by a
   sequence of CL-additions and subdivisions. Each of the CL-additions and subdivisions keeps the condition that every block is a cycle. Then every block of G is
   a cycle.
2. Every proper 8-subgraph has a cut-vertex contained in two cycles, and every θ-
   subgraph is contained in a block. Thus, if every block is a bridge or a cycle, and
   every cut-vertex is contained in at most one cycle, then G has neither a proper
   8-subgraph nor a θ-subgraph; hence, G is a bridged type. Conversely, suppose G
   is a bridged type. By Corollary 4.1, G is constructed from a loop by a sequence of
   LB-additions and subdivisions. Each of the LB-additions and subdivisions keeps
   the condition. Then every block of G is a bridge or a cycle, and every cut-vertex
   of G is contained in at most one cycle.
3. Clearly, every connected graph without cut-vertex contains neither a proper 8-
   subgraph nor a proper manacle-subgraph. Conversely, suppose G is a fused type.
   By Corollary 4.1, G is constructed from a loop by a sequence of CE-additions
   and subdivisions. Because each of the CE-additions and subdivisions does not
   produce a cut-vertex, every block of G has no cut-vertex.

We will include the classification of graph types with rank up to 4.

4.3 Types of graphs and adjacency matrix

Renaming the vertices $v_1, \cdots, v_n$ of G produces a permutation of rows and columns
of A_G.

Proposition 4.1. *Let G be a graph with adjacent matrix A.*

*1. G is disconnected if and only if A can be transformed into a direct sum of two
   square matrices by renaming the vertices. Namely, A is represented by a block
   matrix as follows.*

$$\begin{pmatrix} A_1 & O_{n_1,n_2} \\ O_{n_2,n_1} & A_2 \end{pmatrix} = \begin{pmatrix} A_1 & O_{n_1,n_2} \\ O_{n_2,n_1} & O_{n_2,n_2} \end{pmatrix} + \begin{pmatrix} O_{n_1,n_1} & O_{n_1,n_2} \\ O_{n_2,n_1} & A_2 \end{pmatrix}$$

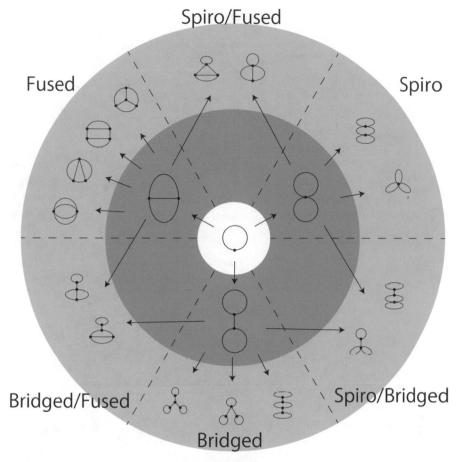

Fig. 4.6 Classification of types of graphs with rank 2 and 3.

Here O_{n_1,n_2} is the $n_1 \times n_2$ zero matrix, in which all elements are zero.

2. *G has a cut-vertex if and only if by renaming the vertices A is represented as follows.*

$$\begin{pmatrix} A_1 & O_{n_1,n_2-1} \\ O_{n_2-1,n_1} & O_{n_2-1,n_2-1} \end{pmatrix} + \begin{pmatrix} O_{n_1-1,n_1-1} & O_{n_1-1,n_2} \\ O_{n_2,n_1-1} & A_2 \end{pmatrix},$$

where A_1 and A_2 are non-zero square matrices. Here n_1 or n_2 can be one (G has a loop in that case).

3. *G has a bridge if and only if, by renaming the vertices, A is represented by a block matrix $\begin{pmatrix} A_1 & J \\ {}^t J & A_2 \end{pmatrix}$, where J is a matrix in which a single element is one and the rest of the elements are zero (a single-entry matrix).*

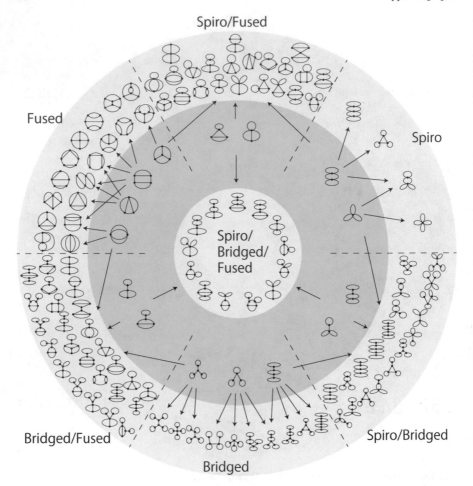

Fig. 4.7 Classification of types of graphs with rank 3 and 4.

Example 4.2. The graph $(5,3^3)_2$ has a cut-vertex.

$$(5,3^3)_2 \quad\quad \begin{pmatrix} 0&2&1&0 \\ 2&0&1&0 \\ 1&1&0&3 \\ 0&0&3&0 \end{pmatrix} = \begin{pmatrix} 0&2&1&0 \\ 2&0&1&0 \\ 1&1&0&0 \\ 0&0&0&0 \end{pmatrix} + \begin{pmatrix} 0&0&0&0 \\ 0&0&0&0 \\ 0&0&0&3 \\ 0&0&3&0 \end{pmatrix}$$

The graph $(4,3^4)_5$ has a bridge.

$$(4,3^4)_5 \quad\quad \begin{pmatrix} 0&2&1&0&0 \\ 2&0&1&0&0 \\ 1&1&0&1&0 \\ 0&0&1&0&3 \\ 0&0&0&3&0 \end{pmatrix}$$

References

1. Y. Tezuka, Acc. Chem. Res., **50**, 2661 (2017).

Chapter 5
Knot theory analysis of polymers

In this chapter, we discuss topological isomers of multicyclic polymers by using knots, links, and spatial graphs. Essential references on knot theory are [1, 2, 3, 4]. Chemistry applications of knot theory and low-dimensional topology are widely discussed in [5, 6].

5.1 Knots, links, and spatial graphs

Knots appear in many places. *Knots* and *links (catenanes)* can be found in closed circular DNA [7, 8, 9] and in proteins[10, 11, 12]. Knot theory has been successfully applied to site-specific DNA recombination [13, 14]. Knots, links (catenanes), as well as *spatial graphs*, are also found in other polymers. These structures change polymer properties, including chirality[5, 6], and the radius of gyration[15]. Because the structures can be observed in polymers via atomic force microscopy, knot theory is important. See also [16, 17, 18, 19, 20, 21, 22, 23, 24, 25, 26, 27, 28] for reports on various molecules having knotted or linked structures.

Knots, links, and spatial graphs yield *topological isomers*. See [29, 30] for a discussion on topological isomers. We will discuss this in Section 6.2. A discussion of knots, links, and spatial graphs is essential for understanding the structure of polymers.

Knots, links, and spatial graphs are also sources of *chirality* in polymers. We will discuss this in Section 5.3.

5.1.1 Knots and links

We start with a mathematical definition of knots and links.

Definition 5.1. A *knot* is a closed circle in 3-dimensional space, i.e., an embedding of S^1 in 3-dimensional space or its image. A *link* (or a *catenane*) is a disjointed

K. Shimokawa et al., *Topology of Polymers*, SpringerBriefs in the Mathematics of Materials 4, https://doi.org/10.1007/978-4-431-56888-9_5

union of knots. Usually, we consider knots and links in 3-dimensional Euclidean space $\mathbb{R}^3$.

Definition 5.2. Two knots (resp. links) k_1 and k_2 are *equivalent* if k_1 can be continuously deformed into k_2 without cutting and transpassing. Mathematically, k_1 and k_2 in $\mathbb{R}^3$ are equivalent if they are *ambient isotopic*, *i.e.*, there is a continuous map $H : \mathbb{R}^3 \times I \to \mathbb{R}^3$ such that $h_0 : \mathbb{R}^3 \to \mathbb{R}^3$ is the identity map of $\mathbb{R}^3$ and $h_1(k_1) = k_2$, where $h_t : \mathbb{R}^3 \to \mathbb{R}^3$ is a homomorphism for any $t \in I$ defined by $h_t(x) = H(x,t)$ for $x \in \mathbb{R}^3$.

A knot is *trivial* or *unknotted* if it is equivalent to the unit circle in $\mathbb{R}^2$ (Top leftmost in Fig. 5.1). Otherwise, it is *non-trivial* or *knotted*.

Definition 5.3. Let k be a knot or link in $\mathbb{R}^3$ and $\pi : \mathbb{R}^3 \to \mathbb{R}^2$ the projection such that $\pi(x,y,z) = (x,y)$. Then $\pi(k) \subset \mathbb{R}^2$ is the *projection* of k. We assume that multiple points of π on k are a finite number of double points. A diagram D of k is the projection $\pi(k)$ with information regarding over-strands and under-strands at each crossing. The crossing number $c(D)$ of D is the number of crossings (double points). The crossing number $c(k)$ of k is the minimum of $c(D)$, where D is any diagram of k.

A diagram D representing a non-trivial knot contains at least 3 crossings. However, the crossing number of the Hopf link is 2. Here, we include a table of knots and links with up to 5 crossings. A knot with 3 crossings is a *trefoil knot* and is denoted by 3_1. A knot with 4 crossings is a *figure-eight knot* and is denoted by 4_1.

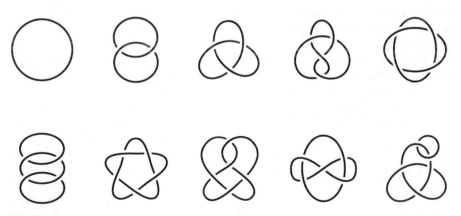

Fig. 5.1 Knots and links up to five crossings. Five are knots, four are two-component links (i.e., links with two connected components), and one is a three-component link. Top from left: unknot (trivial knot) 0_1, Hopf link 2_1^2, trefoil knot 3_1, figure-eight knot 4_1, and $(2,4)$-torus link 4_1^2.

Definition 5.4. An *orientation* of a knot is a choice of direction along a knot. The orientation is often represented by an arrow. An *oriented knot* is a knot with an orientation. An oriented link is a link with orientations for each component. Two oriented links are *equivalent* if they are ambiently isotopic and the isotopy sends the orientation of one link to that of the other.

Definition 5.5. Let $\ell = k_1 \cup k_2$ be an oriented link with two components. At each crossing of a strand of k_1 and k_2, we assign $+1$ or -1 as in Fig. 5.2. The *linking number of* ℓ is half the sum of those numbers.

The linking number is an oriented-link invariant, *i.e.*, two equivalent oriented links have the same linking number. If the linking numbers of two oriented links are different, we can conclude that those oriented links are not equivalent.

Fig. 5.2 We assign $+1$ or -1 at each crossing of a strand of k_1 and a strand of k_2. The oriented link at right has linking number -2. If we change the orientation of one component, the linking number will be changed to $+2$.

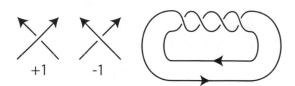

5.1.2 Spatial graphs

A *spatial graph* is a generalization of a knot and a link. Multicyclic polymers in three-dimensional space can be considered spatial graphs.

Definition 5.6. Let G be a graph. A realization K of G in 3-dimensional space, *i.e.*, an embedding of the polyhedron $|C_G|$ of G in $\mathbb{R}^3$, is a *spatial graph*. See Fig. 5.3 for examples.

Definition 5.7. Two spatial graphs k_1 and k_2 are *equivalent* if k_1 can be continuously deformed into k_2, i.e., if k_1 and k_2 are ambiently isotopic.

As in the knot case, for a spatial graph k we define *diagram* and *crossing number* for k. For example, we can show that the crossing numbers of spatial embeddings of graph $K_{3,3}$ in Fig. 5.3 are 1 and 3, respectively.

Definition 5.8. Let $k = f(|C_G|)$ be a spatial embedding of graph G. For a cycle γ of G, $f(|C_\gamma|)$ is a *constituent knot*. For subgraph H of G, $f(|C_H|)$ is a *constituent spatial graph*.

For example, as $K_{3,3}$ has 15 distinct cycles, there are 15 constituent knots in a spatial embedding of $K_{3,3}$.

Example 5.1. Every constituent knot of the spatial graph of $K_{3,3}$ in Fig. 5.3 (left) is an unknot. The set of constituent knots of 3_1 of one in Fig. 5.3 (right) contains a trefoil knot.

From the definition of the equivalence of spatial graphs, we can show the following result.

Proposition 5.1. *Let k_1 and k_2 be two spatial embeddings of graph G. If k_1 and k_2 are equivalent, then the sets of constituent knots of k_1 and k_2 coincide.*

Using this proposition, we can show that two spatial graphs of $K_{3,3}$ in Fig. 5.3 are not equivalent.

Example 5.2. The converse of Proposition 5.1 does not hold in general. For example, every constituent knot of spatial embeddings 0_1 and 5_1 of the θ-graph in Fig. 5.4 is trivial. However, these spatial graphs are not equivalent.

We include tables of spatial graphs. Simon [29] made tables of K_4 and θ spatial graphs . The enumeration of prime embedding types of the θ and manacle (hand-cuff) graphs are given by Moriuchi[31, 32]. In our table, non-prime examples are included.

The enumeration of spatial embeddings of $K_{3,3}$ up to three crossings is given by Fukaguchi[33].

Theorem 5.1. [33] *There are precisely four spatial graphs of $K_{3,3}$ with crossing number 3.*

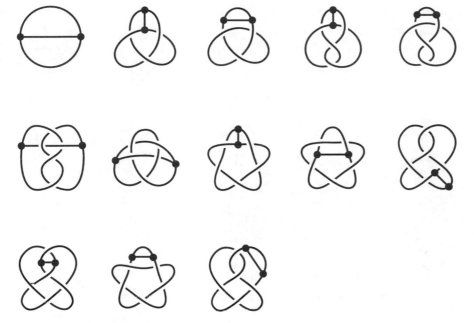

Fig. 5.5 Spatial embeddings of θ graphs up to five crossings.

5.2 Non-planarity

Definition 5.9. A graph G is *planar* if G can be embedded in a plane. Otherwise, it is *non-planar*.

K_5 and $K_{3,3}$ are non-planar graphs. Moreover, the following theorem is known.

Theorem 5.2 (Kuratowski's theorem). *A graph G is non-planar if and only if G has a subgraph homeomorphic to a polyhedron of K_5 or $K_{3,3}$.*

For K_n, Conway and Gordon proved the following theorem.

Theorem 5.3 (Conway-Gordon theorem[34]).

1. *Each spatial embedding of K_7 contains a non-trivial knot.*
2. *Each spatial embedding of K_6 contains a non-trivial link.*

5.3 Chirality

Knots, links, and spatial graph structures induce *topological chirality* in polymers. See [5, 6] for detailed arguments.

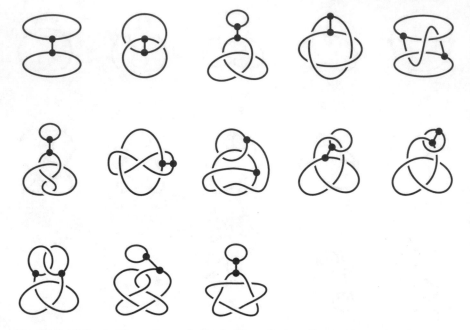

Fig. 5.6 Spatial embeddings of manacle (handcuff) graphs up to five crossings.

Fig. 5.7 Spatial embeddings of the $K_{3,3}$ graph up to three crossings.

Definition 5.10. A spatial graph k is *rigidly chiral* if k cannot be transformed into its mirror image k^* by an orientation preserving the rigid transformation of $\mathbb{R}^3$, *i.e.*, a composition of rotations and parallel translations.

Definition 5.11. A knot, link, or spatial graph k is *topologically chiral* if k is not equivalent to its mirror image k^*. Otherwise, it is *topologically achiral* or *amphichiral*.

Proposition 5.2. *Let k be an oriented knot. Then $V_{k^*}(t) = V_k(t^{-1})$.*

Using Proposition 5.2, we can check the chirality of knots using the Jones polynomial. As $V_{3_1}(t) = t + t^3 - t^4$, we have $V_{3_{1*}}(t) = t^{-1} + t^{-3} - t^{-4}$. It shows that 3_1 is not equivalent to its mirror image 3_1*. Hence 3_1 is chiral. With respect to the relationship of rigid and topological chirality, we can show the following.

Proposition 5.3. *If a spatial graph k is topologically chiral, then it is rigidly chiral.*

For spatial graphs, we can show the following result.

Proposition 5.4. *If a spatial graph k has a topologically chiral knot in its set of constituent knots and it has a mirror image, then k is topologically chiral.*

Using this proposition, we can show that spatial graphs of $K_{3,3}$ with three crossings in Fig. 5.7 is topologically chiral.

References

1. P. Cromwell, Knots and links, Cambridge University Press, Cambridge, 2004.
2. A. Kawauchi ed., A survey of knot theory, Birkhäuser Verlag, Basel, 1996.
3. K. Murasugi, Knot theory and its application, Birkhäuser, Boston, 1996.
4. D. Rolfsen, Knots and links, Mathematics Lecture Series, 7. Publish or Perish, Inc., Houston, TX, 1990.
5. E. Flapan, When topology meets chemistry. A topological look at molecular chirality. Outlooks. Cambridge University Press, Cambridge; Mathematical Association of America, Washington, DC, 2000.
6. E. Flapan, Knots, molecules, and the universe: an introduction to topology. American Mathematical Society, Providence, RI, 2016.
7. A.D. Bates and A. Maxwell, DNA Topology, 2nd ed. (Oxford University Press, Oxford, 2005).
8. J Arsuaga, M Vazquez, S Trigueros, and J Roca, Proc. Nat. Acad. Sci. USA **99**, 5373-5377 (2002).
9. J Arsuaga, M Vazquez, P McGuirk, S Trigueros, and J Roca, Proc. Nat. Acad. Sci. USA **102**, 9165-9169 (2005).
10. J.I. Sulkowska, E.J. Rawdon, K.C. Millett, J.N. Onuchic, and A. Stasiak, Proc. Natl. Acad. Sci. U.S.A. **109**, E1715-E1723 (2012).
11. M. Jamroz, W. Niemyska, E.J. Rawdon, A. Stasiak, K.C. Millett, P. Sulkowski, J.I. Sulkowska, Nucleic Acids Res. **43**, D306-D314 (2014).
12. P. Dabrowski-Tumanski, P. Rubach, D. Goundaroulis, J. Dorier, P. Sulkowski, K.C. Millett, E.J. Rawdon, A. Stasiak, J.I. Sulkowska, Nucleic Acids Res. **47**, D367-D375(2018).
13. C. Ernst and D.W. Sumners, Math. Proc. Camb. Philos. Soc. **108**, 489-515 (1990).
14. K. Shimokawa, K. Ishihara, I. Grainge, D. J. Sherratt, and M. Vazquez, Proc. Nat. Acad. Sci., USA, **110**, 20906 (2013).
15. E. Uehara and T. Deguchi, J. Chem. Phys. **147**, 094901 (2017); doi: 10.1063/1.4996645.
16. C.O. Dietrich-Buchecker, J.-P. Sauvage and J.P. Kintzinger, Tetrahedron Lett. **24**, 5095-5098 (1983).
17. J.-F. Nierengarten, C.O. Dietrich-Buchecker, and J.-P. Sauvage, J. Am. Chem. Soc. **116**, 375-376 (1994).
18. F. Ibukuro, M. Fujita, K. Yamaguchi, and J.-P. Sauvage, J. Am. Chem. Soc.**121**, 11014-11015 (1999).
19. K.S. Chichak. et al., Science **304**, 1308-1312 (2004).
20. R.S. Forgan, J.-P. Sauvage, and J.F. Stoddart, Chem. Rev. **111**, 5434-5464 (2011).
21. D.A. Leigh, R.G. Pritchard, and A.J. Stephens, Nat. Chem. **6**, 978-982 (2014).
22. C.S. Wood, T.K. Ronson, A.M. Belenguer, J.J. Holstein, and J.R. Nitschke, Nat. Chem. **7**, 354-358 (2015).
23. R, Zhu, J. Lübben, B. Dittrich, and G.H. Clever, Angew. Chem. Int. Ed. **54**, 2796-2800 (2015).
24. G. Gil-Ramírez, D.A. Leigh, and A.J. Stephens, Angew. Chem. Int. Ed. **54**, 6110-6150 (2015).
25. C.J. Bruns and J.F. Stoddart, The Nature of the Mechanical Bond: From Molecule to Machines (John Wiley & Sons, Inc., Hoboken, 2016).
26. J.J. Danon, D.A Leigh, S. Pisano, A. Valero, and I.J. Vitorica-Yrezabal, Angew. Chem. Int. Ed. **57**, 13833-13837 (2018).

27. L. Zhang et al., Nat. Chem. **10**, 1083-1088 (2018).
28. T. Sawada, A. Saito, K. Tamiya, K. Shimokawa, Y. Hisada, and M. Fujita, Nat. Comm. **10**, Article number: 921 (2019).
29. J. Simon, in Graph Theory and Topology in Chemistry, King, R. B. and Rouvray, D. H., Eds.; Elsevier: Amsterdam, The Netherlands, 1987.
30. J.C. Dobrowolski, Croat. Chem. Acta **76**, 145-152 (2003).
31. H. Moriuchi, J. Knot Theory Ramifications **18**, 167-197 (2009).
32. H. Moriuchi, OCAMI Studies Vol. 1 (2007), Knot theory for Science Objects, 179-200.
33. S. Fukaguchi, Master thesis, Saitama university (2017).
34. J. H. Conway and C. McA. Gordon, J. Graph Theory **7** 445-453 (1983).

Chapter 6
Topological operations and chemical isomerism of polymers

The control of topological chain properties is essential in biopolymer processes, including DNA transcription and replication promoted by topoisomerase enzymes. Cleavage and re-bonding of DNA chains enables transformation of the chain topologies between a trivial knot (simple ring) and higher knotted constructions, as well as linked counterparts[1, 2]. Recently, topologically remarkable branching/folding structures have been identified in a class of proteins and polypeptides containing S-S bridging of selected cysteine pairs located along the cyclic polypeptide backbone[3, 4]. Their precise spatial 3D structures are crucial for specific biological functions, and enable chemical, thermal, and enzymatic stabilities [3, 4]. Ongoing breakthroughs in synthetic polymer chemistry now allow extensive choices in macromolecular structures beyond linear or randomly branched ones. Synthetic polymer systems with sufficiently long, flexible chains for random coil conformations are compatible with topological geometry conjectures. They are characterized by terminus (chain end) and junction (branching point) numbers as key invariant geometric parameters[5]. In this Chapter, we discuss the geometrical/topological operation of graph-structure polymers, with an emphasis on the relationships between chemical isomerism of topological polymers and geometrical graph transformations. A unique conception of *topological isomerism* is introduced that contrasts with constitutional and stereoisomerism in small molecules. From these considerations, rational and practical synthetic pathways are discussed for complex polymers having cyclic and multicyclic topologies.

6.1 Constitutional, stereo, and topological isomerism in graph-structure polymers

Isomerism has been a fundamental concept in chemistry since Berzelius [6]. Molecules having the same atomic constitution (and molar mass), but different properties, are isomers. The term is derived from the Greek, *isos* (equal) and *meros* (part), and the concept through Kekulé[7] and van't Hoff[8] has continuously provided a deeper understanding of static and dynamic molecular structures. Constitutional (structural) isomers are molecules with distinctive *connectivity* with respect to atoms or atomic groups. Whereas, stereoisomers have indistinguishable *connectivity*, but differ with respect to Euclidian geometric rigidity, i.e., restrictions in bond angle bending and rotation. An increasing number of topologically unique molecules having cyclic[9], knot[10], and catenane [10] forms have been identified in bioresource systems and in synthetic polymer chemistry.

In this section, constitutional isomerism in polymers with flexible chain segment components are discussed[11] with regard to constitutional isomerism of small molecules with Euclidian geometric properties. Synthetic processes for polymeric constitutional isomers having cyclic units are discussed with regard to the topological analysis of graph constructions.

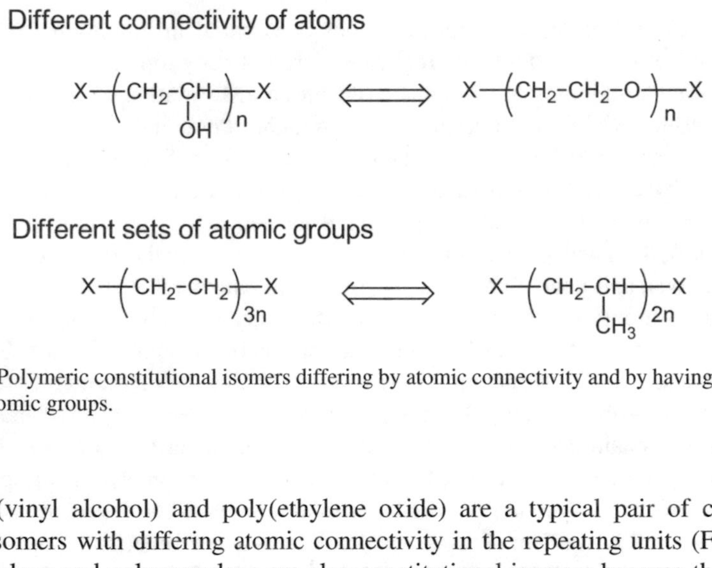

Different connectivity of atoms

$$X-\left(CH_2-\underset{\underset{OH}{|}}{CH}\right)_n-X \quad \Longleftrightarrow \quad X-\left(CH_2-CH_2-O\right)_n-X$$

Different sets of atomic groups

$$X-\left(CH_2-CH_2\right)_{3n}-X \quad \Longleftrightarrow \quad X-\left(CH_2-\underset{\underset{CH_3}{|}}{CH}\right)_{2n}-X$$

Fig. 6.1 Polymeric constitutional isomers differing by atomic connectivity and by having different sets of atomic groups.

Poly(vinyl alcohol) and poly(ethylene oxide) are a typical pair of constitutional isomers with differing atomic connectivity in the repeating units (Fig. 6.1). Polyethylene and polypropylene are also constitutional isomers because they have different sets of atomic groups (Fig. 6.1). These constitutional isomers are identified by formulas of atomic groups.

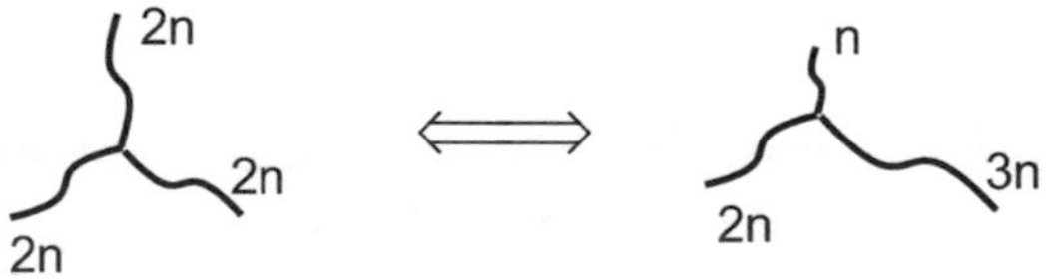

Fig. 6.2 Polymeric constitutional isomers with identical connectivities via different sets of chains.

Another type of constitutional isomerism involves a pair of star polymers having the same number of arms and total arm lengths, but having different sets of arm length compositions (Fig. 6.2). These isomer structures can be expressed simply by graph construction, without the use of chemical formulas. The molar mass then corresponds to the total chain length of the polymer graph, while junctions and terminus units can be excluded. A pair of molecular graphs are interconvertible by the hypothetical continuous extension/contraction of the component chain length, which relates to a basic topological transformation.

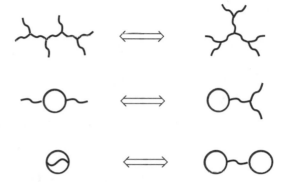

Fig. 6.3 Polymeric constitutional isomers with different connectivities of identical sets of chains.

Another topological class involves pairs of dendritic and comb-type branched polymers, monocyclic polymers with two-tail tadpole and y-tail tadpole forms, and dicyclic polymers with θ and manacle-forms (Fig. 6.3). Interconversion can be completed via chain-breaking at two or more positions, followed by chain-rearrangement/recombination, but not via extension/contraction of chain length components. These constitutional isomers are *topologically distinctive* in contrast to the *topologically equivalent* constitutional isomers of star polymers.

The synthesis of *topologically distinctive* constitutional polymeric isomers occurs through the prescribed linking chemistry with identical sets of prescribed polymer precursors. For example, the isomer pair $^1A_5(2,2)$ and $^1A_6(2,2)$ are obtained simultaneously by using a set of two bifunctional and two monofunctional polymer precursors with two trifunctional end-linking reagents (Fig. 6.4). Such constitutional isomer pairs also occur in ^0A main-class topologies, such $^0A_9(6,3)$ and $^0A_{10}(6,4)$ in Table 3.1, as well as in the ^1A main-class, such as $^1A_6(3,3)$, $^1A_7(3,3)$, and $^1A_8(3,3)$ in Table 3.2.

The synthesis of various polymer topologies, including tadpole forms and dicyclic isomeric pairs of θ and manacle-forms, has been performed via electrostatic self-assembly and covalent fixation (ESA-CF)[11, 12]. In ESA-CF, linear telechelics are used that have moderately strained cyclic ammonium cations accompanying nucleophilic carboxylate counter anions at designated positions. For example, a tadpole polymer has been synthesized via the combination of one difunctional and one monofunctional telechelic precursor carrying one trifunctional counter anion. Dynamic reshuffling in ESA takes place under dilution to maintain the charge balance. (Fig. 6.5) A subsequent covalent conversion by a ring-opening reaction of cyclic ammonium salt groups by carboxylate counter anions could produce tadpole polymers [13].

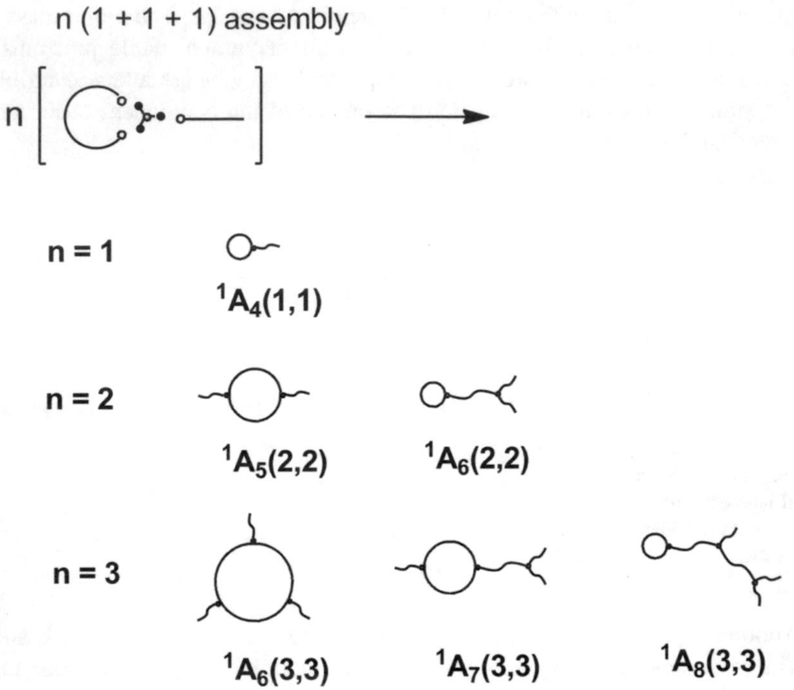
n (1 + 1 + 1) assembly
n
n = 1
1A4(1,1)
n = 2
1A5(2,2)
1A6(2,2)
n = 3
1A6(3,3)
1A7(3,3)
1A8(3,3)

Fig. 6.4 Polymeric constitutional isomers produced from a linear precursor set for tadpole-form synthesis.

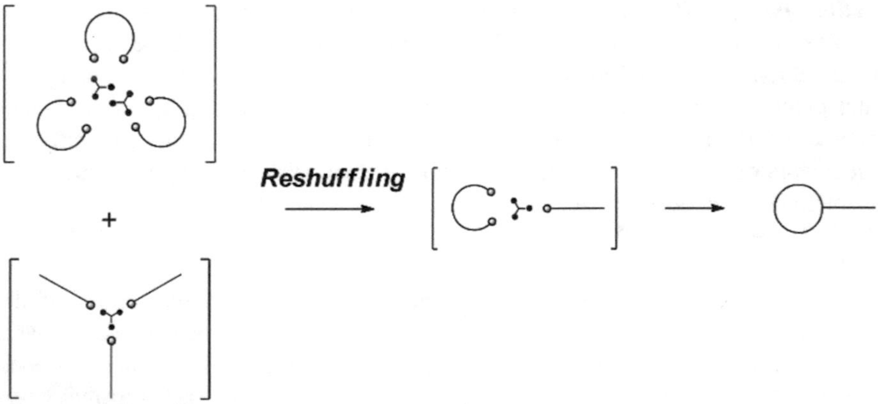
Reshuffling
+

Fig. 6.5 ESA-CF synthesis of a tadpole polymer topology by reshuffling linear polymer assemblies.

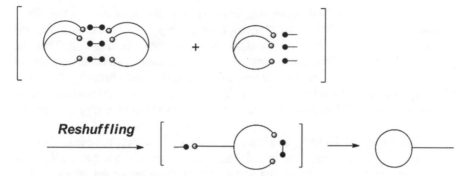

Fig. 6.6 ESA-CF synthesis of a tadpole polymer topology by reshuffling star polymer assemblies.

An alternative ESA-CF synthesis of tadpole polymer topologies has been performed by using a star telechelic precursor having a mixture of mono- and dicarboxylates. The dynamic selection of the two types of carboxylates occurs by reshuffling to selectively form the thermodynamically favored product of star precursors with one mono- and one dicarboxylate counter anion [13] (Fig. 6.6). The subsequent covalent conversion by heating proceeded homogeneously under dilution to produce tadpole polymers.

6.2 Topological analysis of isomerism in monocyclic polymers

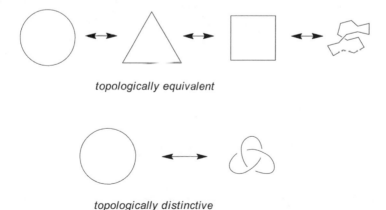

Fig. 6.7 Polymeric stereoisomers via loop construction.

As shown in Fig. 6.7, an open ring polymer is a stereoisomer of a triangle, a square, or a randomly folded counterpart, because they are all interconvertible via conceptual continuous chain-deformation and intact total chain length. Thus, no chain-breaking is involved in this geometrical transformation. In contrast, a pair of ring and knot polymers are interconvertible, but require chain-breaking (Fig. 6.7). More specifically, chain-breaking at a SINGLE location and subsequent chain-rearrangement/recombination completes the interconversion of a ring/knot isomer pair.

A simple ring (trivial knot) and knot (typically a trefoil) pair have been listed as *topological stereoisomers* in the literature. This is because of a topological geometry theorem in which a simple ring and a knot are interconvertible when handled in four-dimensional space. However, this is counterintuitive in three-dimensional chemistry. The ring-knot transformation by chain-breaking at a SINGLE location is intrinsically different from that applied to constitutional isomers that requires chain-breaking at TWO positions, at least, to complete the interconversion.

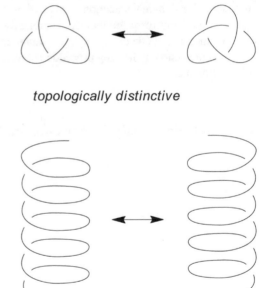

topologically distinctive

topologically equivalent

Fig. 6.8 Enantiomeric stereoisomers of trefoil knots and helix polymers. Two trefoil knots are not topologically equivalent [not ambient isotopic (See Definition 5.2)]; whereas, two helices are topologically equivalent.

A trefoil is a chiral graph. Right-handed and left-handed trefoil knot polymeric isomers have been synthesized and resolved as two chiral products[10]. They are interconvertible only via chain-breaking at a SINGLE location, in contrast to classical enantiomeric isomers with molecular graphs that are interconverted via conceptual continuous chain-deformation (Fig. 6.8). Thus, right- and left-handed knot polymers are *topologically distinctive*, while right- and left-handed open-helix chain molecules are *topologically equivalent* (Fig. 6.8).

6.3 Topological analysis of isomerism in dicyclic polymers

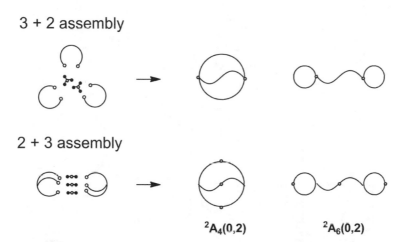

Fig. 6.9 Dicyclic polymeric constitutional isomers from assemblies of linear and star precursors.

The θ- and manacle-shaped polymeric constitutional isomers, $^2A_4(0,2)$ and $^2A_6(0,2)$, respectively, are topologically distinctive constitutional isomers. They are obtained simultaneously via three bifunctional, linear telechelic precursors and two trifunctional linking reagents, or, alternatively, two three-armed telechelic precursors and three bifunctional linking reagents, as seen in Fig. 6.9. The isomers are interconvertible by chain-breaking at TWO positions, followed by chain-rearrangement/recombination.

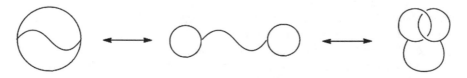

Fig. 6.10 Constitutional and stereoisomerism in θ-, manacle- and pretzelane-form polymers.

An isomer having a pretzelane-form is produced through the catenation of two ring units of the manacle construction (Fig. 6.10). The manacle- and pretzelane-form isomers are interconvertible by chain-breaking at a SINGLE location. Thus, they are classified as topologically distinctive stereoisomers. In topological geometry, a pretzel transformation refers to interconversion between manacle- and pretzelane-forms through unrestricted deformation of chain components involving a figure-8-shaped intermediate; the number of junction points is not retained.

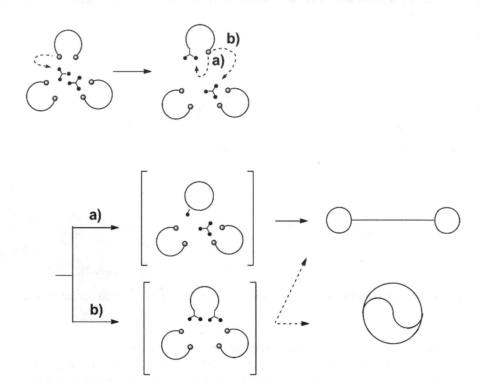

Fig. 6.11 ESA-CF synthesis of dicyclic polymeric topological isomers by the assembly of linear precursors.

The synthesis of a pair of dicyclic θ- and manacle-form polymeric isomers has been performed with ESA-CF (Fig. 6.11)[12, 14]. ESA of three units of a bifunctional telechelic precursor and two units of trifunctional carboxylates was prepared and heated under dilution to transform the ionic interaction into a robust covalent linkage. The random combination of cations and anions within the polymeric self-assembly should result in θ- and manacle-form isomer products in a ratio of 2 : 3. This was comparable to the experiment using chromatography.

An alternative ESA of two units of a three-armed star precursor, associated with three units of difunctional counter anions, has produced a pair of dicyclic, θ-shaped and manacle-shaped polymeric isomers (Fig. 6.12)[15]. The θ and manacle-forms had a ratio of 2 : 3 that was confirmed with chromatography, where the larger hy-

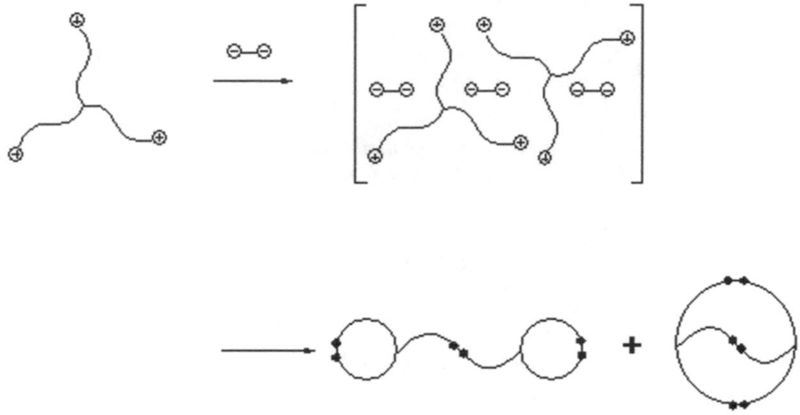

Fig. 6.12 ESA-CF synthesis of dicyclic polymeric topological isomers by assembly of star precursors.

drodynamic volume component was assigned to the manacle-shaped isomer and the smaller hydrodynamic volume to the θ counterpart.

6.4 Topological operation vs. chemical transformation of polymer molecules

In this section, we discuss some features of topological operations in relation to chemical transformations of polymer molecules in practice. First, the isomerization/transformation process between manacle- and pretzelane-form polymeric isomers is compared with the relevant process for a catenane (more precisely [2]catenane, or Hopf link in topology terms) and two separated rings (Fig. 6.13). A [2]catenane is convertible into a simple ring with a double chain length by chain-breaking at TWO positions (ONE each in two ring units), followed by chain-rearrangement/recombination. This exactly conforms to the criterion of constitutional isomers. Whereas, the transformation into two separate rings is completed by chain-breaking at a SINGLE position and chain-rearrangement/recombination. This transformation process is analogous to isomerization between a ring and a trefoil knot, or that between dicyclic manacle- and pretzelane-forms. Chemical isomers are molecules that have the same molar mass, which correspond to total chain lengths in graph construction in the case of polymer substrates. Thus, a [2]catenane molecule with two ring components, each having 30 methylene units, is an isomer of a simple ring molecule having 60 methylene units. More importantly, that are not chemical isomers, but products of two ring molecules having 30 methylene units, just as a cyclo-

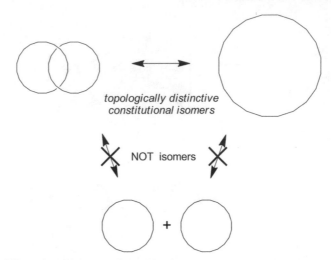

Fig. 6.13 Isomeric relationship between ring and catenane polymers. Two catenanes are distinguishable using linking numbers defined in Definition 5.5

hexane molecule is not an isomer of TWO cyclopropane molecules. However, the dicyclic manacle- and pretzelane-forms are covalently linked products and chemical isomers.

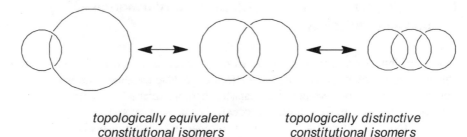

Fig. 6.14 Isomeric relationships between catenane polymers having different ring components.

A large single ring and a series of [n]catenanes are classified as *topologically distinctive* constitutional isomers because each pair is capable of interconversion by chain-breaking at TWO positions, at least, followed by chain-rearrangement/recombination (Fig. 6.14). Whereas, a pair of [2]catenane molecules with two different ring sizes are classified as *topologically equivalent* constitutional isomers, similar to the star polymers with different sets of chain components (Fig. 6.2). The synthesis of polymeric catenanes and knots with long, flexible polymer components has been a formidable challenge, and could provide a unique opportunity to elucidate topology effects on polymer properties[11]. End-linking in polymer cyclization requires a dilution condition to promote intramolecular over intermolecular chain combination. In addition, long, flexible polymer chains tend to coil randomly, contracting the conformation and circumventing threading by another polymer chain, particularly in dilution. Therefore, the preparation of polymeric catenanes usually has less than 1% isolated yields.

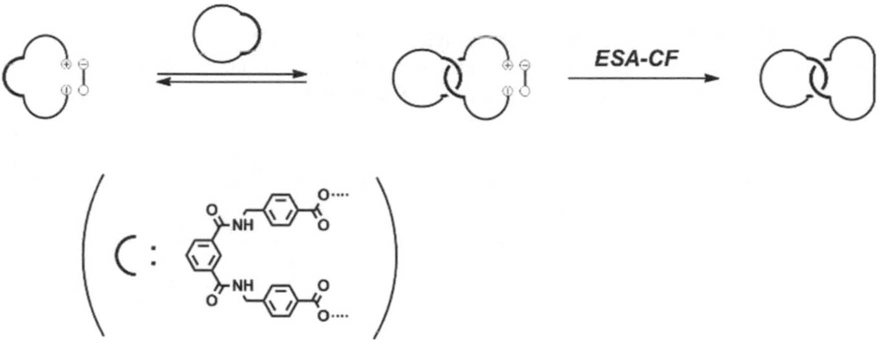

Fig. 6.15 ESA-CF synthesis of a polymer catenane in conjunction with hydrogen-bonding interactions.

ESA-CF has produced polymeric catenanes by taking advantage of attractive interactions of the two polymer chains (Fig. 6.15)[16]. The hydrogen-bonding motif of an isophthaloylbenzylic amide group was introduced to a telechelic precursor for ESA-CF polymer cyclization to form a self-complementary, intertwined chain pair. Two different cyclic polymers with 150-atom ring sizes were prepared as components for a polymeric [2]catenane. The polymer [2]catenane product of two chemically distinct cyclic polymer components was confirmed with mass spectrometry, which can distinguish the catenane product from the simple larger ring. The polymeric catenane was thus isolated in a significantly improved yield of 5.7-7.1% relative to the trace yield (0.1%) in control experiments.

References

1. K. Shimokawa, K. Ishihara, I. Grainge, D. J. Sherratt, and M. Vazquez, Proc. Nat. Acad. Sci. USA, **110**, 20906 (2013).
2. N. C. Seeman, Annu. Rev. Biochem., **79**, 65 (2010).
3. S. J. de Veer, J. Weidmann, and D. Craik, Acc. Chem. Res,. **50**, 1557 (2017).
4. P. Dabrowski-Tumanski and J. I. Sulkowska, Proc. Nat. Acad. Sci. USA **114**, 3415 (2017).
5. Y. Tezuka, Ed., Topological Polymer Chemistry: Progress of cyclic polymers in syntheses, properties and functions (World Scientific, Singapore, 2013).
6. J. J. Berzelius, Pogg. Ann., **19**, 305 (1830).
7. F. A. Kekulé, Ann. Chem., **106**, 129 (1858).

8. J. H. van't Hoff, Archives néerlandaises des sciences exactes et naturelles, **9**, 445 (1874).
9. J. A. Semlyen, Ed., Cyclic Polymers, 2nd ed. (Kluwer, Dordrecht, 2000).
10. J.-P. Sauvage and C. Dietrich-Buchecker, Eds., Molecular Catenanes, Rotaxanes and Knots (Wiley-VCH, Weinheim, 1999).
11. Y. Tezuka, Ed., Topological Polymer Chemistry: Progress of cyclic polymers in syntheses, properties and functions (World Scientific, Singapore, 2013).
12. H. Oike, H. Imaizumi, T. Mouri, Y. Yoshioka, A. Uchibori, and Y. Tezuka, J. Am. Chem. Soc., **122**, 9592 (2000).
13. H. Oike, A. Uchibori, A. Tsuchitani, H.-K. Kim, and Y. Tezuka, Macromolecules, **37**, 7595 (2004).
14. Y. Tezuka, A. Tsuchitani, and H. Oike, Polym. Int., **52**, 1579 (2003).
15. Y. Tezuka, A. Tsuchitani, and H. Oike, Macromol. Rapid Commun., **25**, 1531 (2004).
16. K. Ishikawa, T. Yamamoto, M. Asakawa, and Y. Tezuka, Macromolecules, **43**, 168 (2010).

Chapter 7
Topological polymer chemistry and graph-structure construction

There are abundant examples in which the form of objects dictates their functions and properties at all dimensions and scales. In polymer chemistry and materials science, macromolecular structures have mostly been limited to linear or randomly branched forms. However, a variety of precisely controlled polymer topologies have been synthesized using intriguing techniques[1]. In particular, polymers with cyclic and multicyclic topologies have been made with unprecedented structural precision, and with qualities verified by new spectroscopic and chromatographic techniques[2].

In this chapter, we discuss progress in the synthesis of complex polymer topologies, particularly cyclic and multicyclic molecules. A wide variety of complex multicyclic polymers having *spiro*, *bridged*, *fused*, and *hybrid* forms have been synthesized by ESA-CF in conjunction with recently introduced linking chemistries, such as alkyne-azide addition (click) and olefin metathesis (clip) reactions. A showcase example of *topological polymer chemistry* is the synthesis of a macromolecular $K_{3,3}$ graph topology that will be described below.

The wide range of polymers has inspired modeling and simulations to reveal topological effects. Computational techniques have been able to examine increasingly complex and larger systems [3]. These developments have encouraged experimental verification of the static/dynamic properties unique to cyclic and multicyclic polymers[2]. The properties and functions based on form, i.e. topology, of polymers and polymer design have now been extensively explored[4].

7.1 A ring family tree

All forms of unknotted cyclic topologies up to tricyclic are collected in a "ring family tree" in Fig. 7.1, and are classified into spiro, bridged, fused, and hybrid forms[2, 5]. See also the theoretical discussion in Chapter 4, as well as Fig. 4.6 and Fig. 4.7.

K. Shimokawa et al., *Topology of Polymers*, SpringerBriefs in the Mathematics of Materials 4, https://doi.org/10.1007/978-4-431-56888-9_7

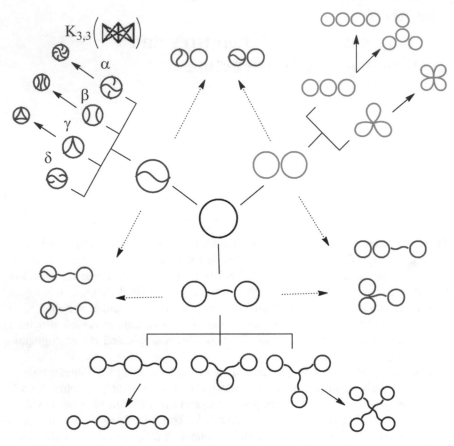

Fig. 7.1 A ring family tree up to tricyclic constructions of *spiro-*, *bridged-*, *fused-*, and *hybrid*-form topologies.

All three dicyclic polymers of eight (*spiro*), manacle (*bridged*), and θ (*fused*) forms, and most tricyclic topologies in *spiro*, *bridged*, *fused*, and *hybrid* forms have been synthesized via ESA-CF in conjunction with covalent linking. The latter includes alkyne-azide *cross-coupling* (*click*), reactions and metathesis *homo-coupling* (*clip*) reactions[5]. ESA-CF has also been used to produce a series of single-cyclic and multicyclic polymer precursors (*kyklo*-telechelics) as large as 300-member atomic rings for subsequent formation of complex multicyclic polymer topologies[6]. These topological polymers are considered important models for experimental investigations of topological effects on basic properties, and for computational modeling/simulation. Eventually, unprecedented polymer design principles based on topologies will be discovered.

7.2 Synthesizing multicyclic polymers with *spiro*- and *bridged*-topologies

A *spiro*-dicyclic, 8-shaped polymer was obtained by ESA-CF with two units of linear telechelic precursor and one unit of tetra-carboxylate counter anion[7]. Alternatively, a metathesis condensation (*clip*) was used with various polymer precursors, including cyclic units (*kyklo*-telechelics), such as a four-armed star telechelic prepolymer with allyloxy end groups, a twin-tailed tadpole with two allyloxy groups at tail-end positions, and a ring polymer with two allyloxy groups at opposite positions[8].

A trefoil *spiro*-tricyclic was obtained by ESA-CF from three units of a linear telechelic precursor with a hexa-functional counter anion[7]. Another tandem form of a *spiro*-tricyclic polymer was synthesized by *click* linking of two complementary *kyklo*-telechelics. One had two alkyne groups at opposite positions of the ring unit, and another had an azide group; both were obtained with ESA-CF(Fig. 7.2)[9].

A *spiro*-tetracyclic polymer having a quatrefoil form was produced by alkyne-azide *click* linking by using a single-cyclic precursor with an azide group via ESA-CF, with a tetra-functional pentaerythritol alkyne derivative (Fig. 7.2)[10]. Another tandem *spiro*-tetracyclic polymer was synthesized by *click* coupling of an 8-shaped *kyklo*-telechelic, with two alkyne groups at opposite positions of the two ring unit, and a single-cyclic precursor having an azide group (Fig. 7.2)[9]. Penta- and heptacyclic *spiro*-form polymers were synthesized via repetitive alkyne-azide *click* linking, using prescribed *kyklo*-telechelics with protected alkyne groups (Fig. 7.2)[11].

A *bridged*-dicyclic, manacle-shaped polymer was obtained together with an isomeric *fused-dicyclic*, θ-shaped product via ESA-CF by using three units of a linear bifunctional precursor with two units of trifunctional carboxylate counter anions. Alternatively, two units of a star-shaped trifunctional precursor with three units of bifunctional carboxylate counter anions could be used, as detailed in Chapter 6[7, 12]. This pair of dicyclic polymeric topological isomers was also formed via a double metathesis *clip* process, using an H-shaped polymer precursor with allyloxy end groups on each chain[13].

Selective synthesis of the manacle-topology polymer was achieved through *click* linking of a bifunctional linear precursor having azide groups and *kyklo*-telechelics with an alkyne group via ESA-CF [9]. By a similar *click* coupling procedure, a *bridged*-tricyclic, three-way, paddle-shaped polymer was obtained from a star-shaped trifunctional precursor having azide groups and *kyklo*-telechelics having alkyne groups (Fig. 7.2)[9]. A *bridged*-type, tetracyclic polymer having a four-way paddle-form was synthesized by *click* linking. It used *kyklo*-telechelics with an azide group and a four-armed star polymer precursor with alkyne end groups (Fig. 7.2)[10].

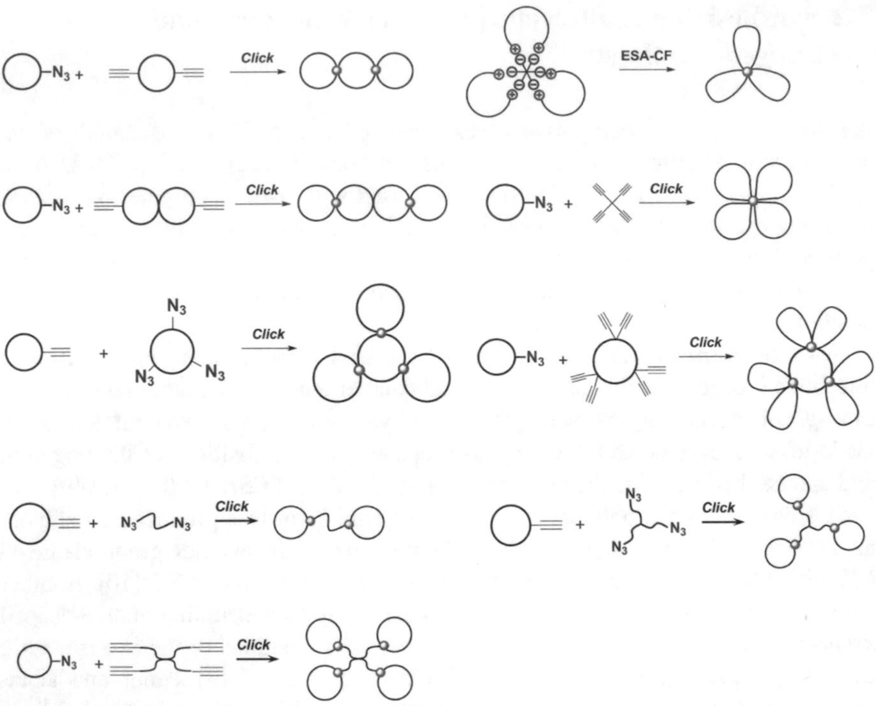

Fig. 7.2 Synthesis of *spiro-* and *bridged-*form multicyclic polymers via ESA-CF and the click method.

7.3 Synthesizing multicyclic polymers with *fused-* and *hybrid*-topologies

A singly-*fused* dicyclic, θ-shaped polymer was obtained via ESA-CF by using a three-armed star precursor with a tricarboxylate counter anion [7, 14]. Alternatively, a θ-shaped polymer was formed together with a manacle-shaped isomer either through ESA-CF or by the metathesis *clip* process, as described above.

A class of doubly-*fused* tricyclic polymer topologies includes α-, β-, γ-, and δ-graph constructions. A δ-graph polymer was first produced by a combination of ESA-CF with the metathesis *clip* process (Fig. 7.3)[15]. An 8-shaped *kyklo*-telechelic with two allyl groups at opposite positions of the two ring units was prepared via ESA-CF using a *kentro*-telechelic (center-functional) polymer precursor with an allyl group at the center of the chain. The subsequent metathesis *clip* condensation with a Grubbs catalyst under dilution allowed folding of the 8-form precursor into a δ-graph polymer product in a good isolated yield. Other doubly-*fused* tricyclic forms of γ-graph and β-graph polymers were synthesized with tandem alkyne-azide *click* and olefin metathesis *clip* reactions in conjunction with ESA-CF(Fig. 7.3)[16]. Thus, a *bridged*-dicyclic (manacle) *kyklo*-telechelic and another counterpart having additional two outward-branching segments, both with two ally-loxy groups at the opposite positions of the ring units or at the branch chain ends, were prepared and subjected to the intramolecular metathesis *clip* reaction under dilution by repeated addition of the Grubbs catalyst. The synthesized γ-graph and β-graph polymers were then isolated by means of size-exclusion chromatography (SEC)[16].

A triply-*fused* tetracyclic polymer with an unfolded tetrahedron graph construction and a p4m symmetry (a D_4 graph[17]), and a quadruply-*fused* pentacyclic polymer in the "shippo" form[18]were obtained via a tandem alkyne-azide *click* and olefin metathesis *clip* reaction with ESA-CF(Fig. 7.3). Thus, *spiro*-type tandem tri- and tetracyclic *kyklo*-telechelics having two allyloxy groups at the opposite positions of the three and four ring units, were first obtained by *click* linking using single-cyclic and dicyclic telechelic precursors (*kyklo*-telechelics) with complementary reactive groups at designated positions in the cyclic and dicyclic units (Fig. 7.3). The subsequent metathesis *clip* reaction under dilution induced folding of the tri- and tetracyclic precursors to yield triply-*fused* tetracyclic and quadruply-*fused* pentacyclic polymer products, respectively. (Fig. 7.3) The multicyclic polymer products were isolated via SEC in good yields[18].

Finally, a class of *hybrid-multicyclic* polymer constructions of an elementary dicyclic unit of either *theta-*, *eight-*, or *manacle*-forms has been produced with a *click* polymer-linking protocol in conjunction with ESA-CF [19]. (Fig. 7.3) A *theta*-shaped *kyklo*-telechelic with an alkyne group at the junction position was prepared as a key precursor by ESA-CF. In addition, a *kyklo*-telechelic having an azide group, as well as a linear-cyclic (tadpole) and a linear-dicyclic (twin-head tadpole) *kyklo*-telechelic, both with azide groups at the tail-end position, were prepared through the tandem *click* reaction of the respective single-cyclic and dicyclic, 8-shaped *kyklo*-telechelics having an alkyne group with a linear asymmetric telechelic, having azide and hydroxyl groups, followed by esterification with 4-azidobenzoic acid.

They were subsequently subjected to the synthesis of a variety of *hybrid*-tricyclic polymer topologies composed of dicyclic (*theta*- or *eight*-shaped) and monocyclic (simple ring or tadpole-shaped) units, and a *hybrid*-tetracyclic topology consisting of *theta*, *eight*, and *manacle* dicyclic units. In addition, double-em eight and double-*theta* topologies have been synthesized through *click*-linking of a pair of *kyklo*-telechelic precursors (Fig. 7.3). A *bridged*-type hexacyclic, double trefoil polymer was synthesized by a click coupling reaction of a core functional trefoil polymer precursor having an alkyne group and a linear precursor having azide end groups[20].

7.4 Constructing a macromolecular $K_{3,3}$ graph topology

A $K_{3,3}$ graph construction is a topologically intriguing non-planer construction that cannot be embedded in a plane in such a way that its edges intersect only at their endpoints (Fig. 7.4). See Theorem 5.2. This non-planarity is also relevant to a Hopf link (2-catenane), with an inherent crossing number of two, and a trefoil knot, with

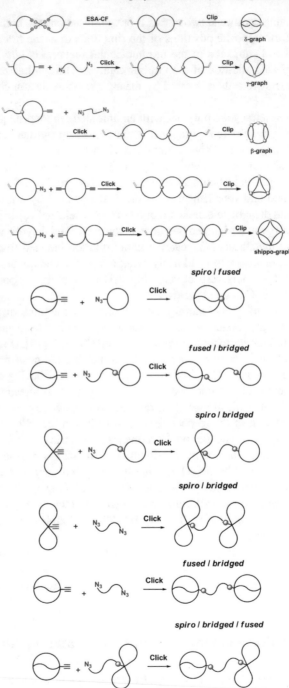

Fig. 7.3 Synthesis of multicyclic polymers having *fused-* and *hybrid-*forms via combined ESA-CF, click, and clip methods.

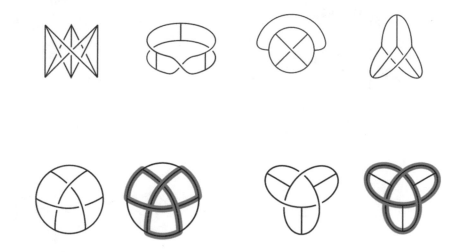

Fig. 7.4 Graph presentations of a $K_{3,3}$ construction (above) and its knot form (below).

a crossing number of three. Therefore, a $K_{3,3}$ graph having a crossing number of 1 is in the class of non-planar graph constructions. The $K_{3,3}$ graph is drawn on a torus surface, avoiding edge intersections[21].

The $K_{3,3}$ graph topology has recently been identified in cyclic polypeptides (cyclotides) produced by intramolecular S-S bridging of cysteine residues[5]. The programmed folding structures are crucial for their extraordinary stability and bioactivity[22]. Programmed folding of linear polymers into designated multicyclic forms is still a formidable challenge in polymer chemistry. It is, therefore, important to explore synthetic approaches to construct topologies known in graph theory and to reveal how polymer topology can direct fundamental properties of flexible polymer molecules.

Based on the geometrical analysis of constitutional isomers obtained from a three-armed star polymer precursor, as seen in Fig. 7.5, the θ-form of $^2A_4(0,2)$ was synthesized with a trifunctional star polymer precursor and a trifunctional end-linking reagent. The two sets of this precursor combination results in the uniquely tricyclic β-graph construction of $^3A_6(0,4)$. The $K_{3,3}$ graph of $^4A_6(0,6)$ is produced by three sets of this particular precursor combination, together with two tetracyclic polymeric isomers in a ladder form and a triply connected tadpole form.

From these geometrical considerations, the synthesis of a macromolecular $K_{3,3}$ graph polymer having uniformly sized edge components of eicosanediol (C_{20}) segments has been achieved by ESA-CF. Thus, a uniformly sized dendritic polymer precursor having six cyclic ammonium salt end groups, with two units of trifunctional carboxylate counter anions, was prepared and subsequently subjected to covalent conversion by the ring-opening reaction of cyclic ammonium salt groups at an elevated temperature under dilution[23]. The $K_{3,3}$ graph product was isolated by

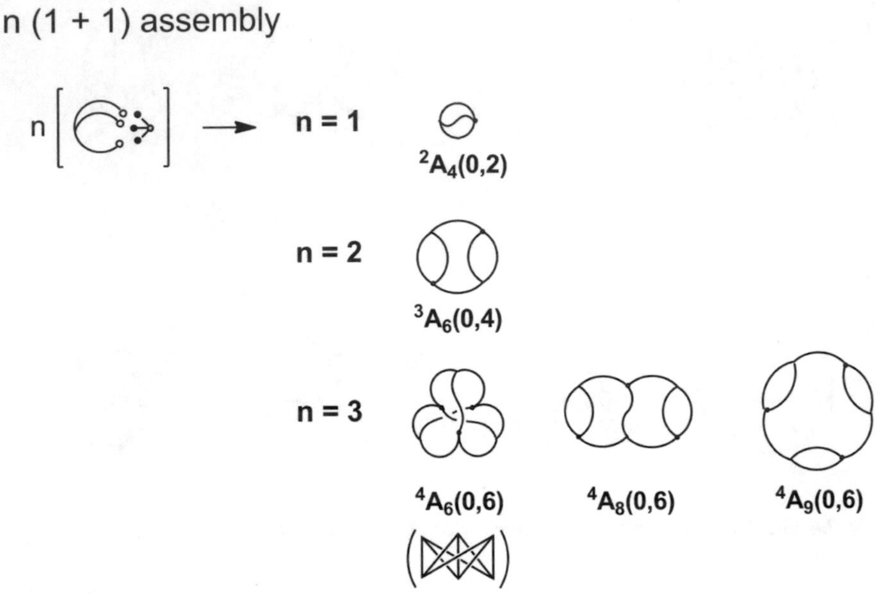

n (1 + 1) assembly
n
n = 1
$^2A_4(0,2)$
n = 2
$^3A_6(0,4)$
n = 3
$^4A_6(0,6)$
$^4A_8(0,6)$
$^4A_9(0,6)$

Fig. 7.5 Polymeric constitutional isomers produced from a star precursor set for a θ-form polymer construction.

2
ESA
CF
A
B
1/5
4/5
C
D
C
1/2
D
1/2
3/5
2/5

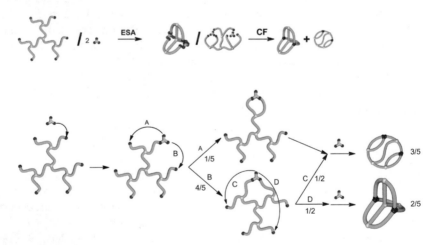

Fig. 7.6 ESA-CF synthesis of a $K_{3,3}$ graph polymer topology and its isomeric ladder form.

recycling SEC, because the hydrodynamic volume of the $K_{3,3}$ graph product was considerably contracted in solution relative to a concurrent isomer having a tetra-cyclic ladder form. (Fig 7.6)

In an analysis of the contractions of a 3D series of multicyclic polymer folding products, the $K_{3,3}$ graph (0.49) polymer product was significantly contracted relative to singly-*fused* θ-shaped (0.63), doubly-*fused* δ-graph (0.61), and γ-graph (0.60) constructions, and was comparable to a triply-*fused* unfolded tetrahedron graph (0.52). This was experimental confirmation that the programmed polymer folding observed in cyclotides could produce unusually compact 3D structures[23, 24].

References

1. N. Hadjichristidis, A. Hirao, Y. Tezuka, and F. Du Prez, Eds., Complex Macromolecular Architectures: Synthesis, Characterization, and Self-Assembly, (Wiley, Singapore, 2011).
2. Y. Tezuka, Ed., Topological Polymer Chemistry: Progress of cyclic polymers in syntheses, properties and functions (World Scientific, Singapore, 2013).
3. T. Deguchi, K. Shimokawa, C. Micheletti, Eds., Knots and Polymers: Aspects of topological entanglement in DNA, proteins and graph-shaped polymers, Special Issue in Reactive and Functional Polymers, (Elsevier, 2019).
4. T. Yamamoto, and Y. Tezuka, Soft Matter, **11**, 7458 (2015).
5. Y. Tezuka, Acc. Chem. Res., **50**, 2661 (2017).
6. H. Oike, S. Kobayashi, T. Mouri, and Y. Tezuka, Macromolecules, **34**, 2742 (2001).
7. H. Oike, H. Imaizumi, T. Mouri, Y. Yshioka, A. Uchibori, and Y. Tezuka, J. Am. Chem. Soc., **122**, 9592 (2000).
8. Y. Tezuka, R. Komiya, and M. Washizuka, Macromolecules, **36**, 12 (2003).
9. N. Sugai, H. Heguri, K. Ohta, Q. Meng, T. Yamamoto, and Y. Tezuka, J. Am. Chem. Soc., **132**, 14790 (2010).
10. Y. S. Ko, T. Yamamoto, and Y. Tezuka, Macromol. Rapid Commun., **35**, 412 (2014).
11. Md. H. Hossain, Z. Jia, and M. J. Monteiro, Macromolecules, **47**, 4955 (2014).
12. Y. Tezuka, A. Tsuchitani, and H. Oike, Macromol. Rapid Commun., **25**, 1531 (2004).
13. Y. Tezuka and F. Ohashi, Macromol Rapid Commun., **26**, 608 (2005).
14. Y. Tezuka, A. Tsuchitani, Y. Yoshioka, and H. Oike, Macromolecules, **36**, 65 (2003).
15. Y. Tezuka, K. Fujiyama, J. Am. Chem. Soc., **127**, 6266 (2005).
16. M. Igari, H. Heguri, T. Yamamoto, and Y. Tezuka, Macromolecules, **46**, 7303 (2013).
17. N. Sugai, H. Heguri, T. Yamamoto, and Y. Tezuka, J. Am. Chem. Soc., **133**, 19694 (2011).
18. H. Heguri, T. Yamamoto, and Y. Tezuka, Angew. Chem., Int. Ed., **54**, 8688 (2015).
19. Y. Tomikawa, T. Yamamoto, and Y. Tezuka, Macromolecules, **49**, 4076 (2016).
20. Y. Tomikawa, H. Fukata, Y. S. Ko, T. Yamamoto, and Y. Tezuka, Macromolecules, **47**, 8214 (2014).
21. E. Flapan, When Topology Meets Chemistry: A Topological Look at Molecular Chirality, Cambridge University Press, Cambridge, UK, (2000).
22. D. J. Craik, Nature Chem., **4**, 600 (2012).
23. T. Suzuki, T. Yamamoto, and Y. Tezuka, J. Am. Chem. Soc., **136**, 10148 (2014).
24. E. Uehara and T. Deguchi, J. Chem. Phys., **145**, 164905 (2016).

Chapter 8
Topology meets polymers: Conclusion and perspectives

In this monograph, we discussed and demonstrated ongoing developments in the unique collaboration of topological geometry and polymer chemistry. We described current *topological polymer chemistry* by highlighting the diverse nature of polymers with respect to both their chemistry and their line constructions. Topological analyses could provide fundamental insights on the principal materials and/or biological properties of polymers based on their segment geometries.

Numerous opportunities are anticipated in *topological polymer chemistry* as a result of these close interactions between mathematics and chemistry. We will soon be able to acquire topological control of static and dynamic polymer properties that rely on geometry conjectures. Sometimes they may be counterintuitive relative to Euclidian geometry. Because a variety of topologically defined but complex polymers have become available, along with the formidable progress in theory and simulations, most topological effects in polymer materials will be uncovered for eventual applications. Thus, we are entering into an exciting era of polymer science and materials engineering based on precision topology design, similar to the "Cambrian explosion period" in the evolution of life. *Topological polymer chemistry*, will certainly contribute to such an exciting new era.

© The Author(s), under exclusive license to Springer Japan KK 2019
K. Shimokawa et al., *Topology of Polymers*, SpringerBriefs in the Mathematics
of Materials 4, https://doi.org/10.1007/978-4-431-56888-9_8

Printed in the United States
By Bookmasters